Algorithmic Trading via AI/Machine Learning with R

Algorithmic Trading via AI/Machine Learning with R aims to demonstrate how algorithmic trading can empower retail traders to compete more effectively in markets long dominated by institutional giants. By translating advanced techniques into practical, systematic strategies, the book shows how automation, disciplined risk management, and data-driven decision making can help individuals filter out market noise, avoid manipulation, and exploit opportunities that once belonged exclusively to large firms.

The book's purpose is to give you a framework where R is not just a statistical environment, but a trading laboratory and execution engine. Every chapter includes reproducible examples you can extend into your own practice and research pipeline. By the end, you will not merely understand algorithmic trading—you will have built, tested, and connected live strategies to market data. At its core, it demonstrates how R—a language renowned for statistical computing—can be transformed into a complete research and execution platform for trading.

This book is aimed at anyone who wants to learn, or use R, for AI/Machine Learning and algorithmic trading. It is also for individuals doing, or interested in doing, securities research and financial systems development and for retail traders who may wish to use R to gain an algorithmic trading edge.

Key Features:

- Follows a clearly defined, pedagogical structure that builds from foundational R tools to full automation and integration with APIs.
- Argues that while retail traders cannot match Wall Street's scale, they can use algorithms to level the playing field—building consistency, resilience, and an edge in a market designed to favor the powerful.
- All the book's scripts can be accessed on the book's GitHub branch.
- The *QuantRoom* YouTube channel (@quantroom) provides video tutorials and scripts that complement the book's content showcasing real-time problem-solving.
- Delivers a more engaging and accessible way to master algorithmic trading using R and the Schwab Trader API.
- The Appendix expands the book's scope beyond R by presenting a side-by-side comparison between the C++ TWS API and the IBrokers R interface, illustrating how high-level R commands map directly to their low-level C++ counterparts.

Algorithmic Trading via AI/Machine Learning with R

Jason Guevara, Ričards Bulavs
and Oskars Linares

CRC Press
Taylor & Francis Group
Boca Raton London New York

CRC Press is an imprint of the
Taylor & Francis Group, an **informa** business
A CHAPMAN & HALL BOOK

Designed Cover Image: Annemarija Linaresa

First edition published 2026
by CRC Press
2385 NW Executive Center Drive, Suite 320, Boca Raton FL 33431

and by CRC Press
4 Park Square, Milton Park, Abingdon, Oxon, OX14 4RN

CRC Press is an imprint of Taylor & Francis Group, LLC

ISBN: 978-1-041-26470-5 (hbk)
ISBN: 978-1-041-26468-2 (pbk)
ISBN: 978-1-003-75045-1 (ebk)

DOI: 10.1201/9781003750451

Typeset in Latin Modern Roman
by KnowledgeWorks Global Ltd.

Publisher's Note: This book has been prepared from camera-ready copy provided by the author.

Access the Support Material: https://github.com/jgvRa/Rbook2.

To all people, whether known or unknown, who have fearlessly stood up for their beliefs for the benefit of humanity.

Contents

Preface

Introduction

For decades, quantitative trading was the guarded domain of Wall Street firms, hedge funds, and institutional desks with armies of PhDs and proprietary infrastructure. The retail trader was left to play catch-up, constrained to chart patterns and broker dashboards. That asymmetry has narrowed. Today, the same statistical methods, data pipelines, and execution interfaces that drive professional trading systems are available to any individual willing to learn.

Our book is written to bridge that gap. It is designed for traders who want to move beyond discretionary chart-reading and into the systematic world of algorithms, models, and automation. At its core, it demonstrates how R—a language renowned for statistical computing—can be transformed into a complete research and execution platform for trading.

The text proceeds in layers:

- **Data Acquisition and Management.** We begin with packages such as quantmod, tidyquant, and IBrokers to fetch live and historical data from Yahoo Finance, FRED, and the Interactive Brokers (TWS API). Readers learn to design reproducible workflows that preserve the full chain of data lineage—from raw imports to transformed signals.

- **Exploratory Analysis and Modeling.** Using dplyr, ggplot2, and PerformanceAnalytics, we explore returns, drawdowns, and factor exposures. Statistical and econometric modeling is extended with lm, glmnet, and Bayesian inference via MCMCpack, while time-series volatility forecasting employs rugarch and tsDyn.

- **Strategy Design and Backtesting.** Strategies are formalized through signal logic—whether simple moving-average crossovers (TTR) or multi-factor frameworks using Fama–French regressions—and validated through PortfolioAnalytics and rolling-window analysis. Performance metrics such as *Sharpe*, *Sortino*, and Conditional VaR (CVaR) anchor the evaluation of robustness and consistency.

- **Machine Learning and AI.** From caret to xgboost and lightgbm, machine learning is applied to market classification, regime detection, and predictive modeling. Reinforcement learning examples demonstrate adaptive agents that respond dynamically to price environments, while clustering and principal component analysis uncover hidden structures in asset behavior.

- **Execution and Infrastructure.** Once signals are validated, the system connects directly to markets through broker APIs. The book shows how to request real-time ticks, send market and limit orders, and deploy bracket orders (entry, stop-loss, and take-profit) using IBrokers. Through these examples, the reader learns to close the research–execution loop: from strategy to automated trading in live or paper environments.

A special chapter is dedicated to *QuantRoom*, the companion YouTube channel and educational platform that extends this text into an applied ecosystem. *QuantRoom* bridges code and practice, offering video walkthroughs, strategy implementations, and live API integrations across Interactive Brokers and Schwab. It transforms static examples into living demonstrations—showing how theoretical models become executable systems. Readers are encouraged to treat *QuantRoom* as both laboratory and classroom: a place to experiment, replicate, and iterate alongside real market data.

The Appendixes expand the book's scope beyond R. It presents a side-by-side comparison between the C++ TWS API and the IBroker's R interface, illustrating how high-level R commands map directly to their low-level C++ counterparts. This comparative structure helps readers understand how the Interactive Brokers API functions at its core—socket communication, event callbacks, message parsing, and order transmission. More importantly, it guides advanced users who wish to transition portions of their R trading systems into C++ for higher performance, lower latency, and multi-threaded execution. By maintaining one-to-one functional mapping, the appendix provides a blueprint for translating strategies from prototype to production, ensuring that speed-critical components can be re-implemented without sacrificing clarity or structure.

This preface is also a challenge: retail traders must stop seeing themselves as outsiders. By systematically applying econometrics, machine learning, and disciplined risk management, you can create strategies that rival professional approaches. The playing field is not equal—institutions still have scale—but technology has eroded their monopoly on tools.

The book's purpose is simple: to give you a framework where R is not just a statistical environment, but a *trading laboratory* and *execution engine*. Every chapter includes reproducible examples you can extend into your own practice and research pipeline. By the end, you will not merely understand algorithmic trading—you will have built, tested, and connected live strategies to market data.

Algorithmic trading is not magic. It is engineering: hypotheses, models, testing, execution, and iteration. With R, the *QuantRoom* ecosystem, and the

C++ appendix as your foundation, you are no longer reacting to the market—you are designing structured, data-driven systems capable of competing within it.

Why is this book important?

Small independent traders, or so-called retail ("home") traders need to understand they are up against the mightiest opponents in the business: Goldman Sachs, BlackRock, Jane Street Capital, JP Morgan Chase, Morgan Stanley, Citadel, etc. For example, Goldman Sachs has not one, but, teams of computational finance engineers (`https://developer.gs.com/discover/gs-quant`). Therefore, retail traders face a daunting challenge: the market is dominated by giants who wield vast capital, proprietary data, and cutting-edge algorithms. These firms don't react to trends—they create them by exploiting liquidity, latency, and privileged information. For the individual trader, trying to compete on the same terms is a losing battle.

But algorithmic trading offers a way to narrow the gap. By systematizing decision-making, filtering out noise, and executing with discipline, retail traders can avoid the traps of overreaction and manipulation. Algorithms help level the field by turning data into consistent rules: they can track institutional flows, detect subtle patterns, and enforce risk management without the emotional swings that institutions often exploit. While retail traders may never match Wall Street's speed or scale, they can fight back with focus, automation, and strategies designed for their niche rather than the giants' game.

A major aim of our book is to demonstrate how algorithmic trading can empower retail traders to compete more effectively in markets long dominated by institutional giants. By translating advanced techniques into practical, systematic strategies, the book shows how automation, disciplined risk management, and data-driven decision-making can help individuals filter out market noise, avoid manipulation, and exploit opportunities that once belonged exclusively to large firms. In short, it argues that while retail traders cannot match Wall Street's scale, they can use algorithms to level the playing field—building consistency, resilience, and an edge in a market designed to favor the powerful.

This book is timely, as it is clear that there is a significant number of retail (or "home") traders currently losing money. This number will undoubtedly increase as the public's access to the stock market rises.

Who will find it useful?

This book is aimed at anyone who wants to learn or use R for AI/Machine Learning and algorithmic trading. It is also aimed at individuals doing, or who may be interested in doing, securities research and financial systems development. In particular, it is aimed at retail ("home") traders who may wish to use R to gain an algorithmic trading edge (*"algo trading edge"*).

What is new?

We (OL) have all of the R-Series books by Chapman & Hall/CRC. In addition, we (OL) also has, all of the Use R! book titles offered by Springer. Of all the Series Titles offered by Chapman & Hall/CRC R Series, we did not find a single title dealing with algorithmic trading using R. In fact, an exhaustive search revealed only one book by APRESS dealing with algorithmic trading using R: *Automated Trading with R: Quantitative Research and Platform Development (Chris Conlan, 2016).* In contrast, there are over 30 titles of books currently available on algorithmic trading using Python.

In Chapter 9 of our book, we introduce a unique pedagogical approach to describe algorithmic trading using R. This approach utilizes explanations, examples, syntax, and YouTube video "walk-throughs" of working code, and provides users with the capability to comment and engage one of us (JG) in dialogue (`https://www.youtube.com/@quantroom`). This approach is unique and new, and it magnifies our book's value. As is usual and customary nowadays, we also provide all the codes used in the book by chapter in the book's Github branch (`https://github.com/jgvRa/Rbook2`).

What background is assumed?

Our book assumes a general-purpose knowledge of the R language and RStudio. However, we provide enough R and RStudio throughout the book and its supplementary materials to make the book stand-alone for entry-level readers. For readers who want to learn more R, there are several books available[1], and advanced readers may want to look at those books[2]. Excellent well-written documentation about the **tidyverse** can be found here: `https://www.tidyverse.org/`.

Are there any books that suffice?

Not really. The only book available is dated, as it was published in 2016. Moreover, it does not cover nearly as many topics as our book. For example, there is no coverage of AI/ML or backtesting using available R packages.

We found the book series by Ernest P. Chan to be the most referenced work on Algorithmic Trading. However, the author uses MATLAB for algorithmic trading system development, not R. A detailed comparison of his series of books with ours follows:

---

[1] `https://www.r-bloggers.com/2022/05/best-books-to-learn-r-programming-2/`

[2] `https://www.rstudio.com/resources/books/`

Quantitative Trading: How to Build Your Own Algorithmic Trading Business (Chan 2009)

Our book expands beyond the original *Quantitative Trading's* focus on infrastructure and business-building by fully integrating AI/ML, R, and C++ interoperability. Our book uses current R packages (mlr3, tidymodels, keras, torch, IBrokers, PortfolioAnalytics) rather than older MATLAB centric or Excel-based workflows. Our book has a stronger technical depth relevance because of its end-to-end reproducible pipelines — data acquisition, wrangling, modeling, backtesting, and live execution — all within R. Furthermore, its appendix directly maps C++ TWS API functions to IBrokers commands, giving a unique cross-language bridge to high-performance trading system design.

Algorithmic Trading: Winning Strategies and Their Rationale (Chan 2013)

Again, our book demonstrates clear superiority — both in scope and scholarly execution. *Algorithmic Trading* focuses narrowly on software selection and workflow setup (e.g., brokers, APIs, data feeds, automation scripts in MATLAB). Its technical content is fragmented — mostly tool descriptions and interface explanations.

Our book expands the theme into a complete quantitative research–to–execution pipeline: Data sourcing, feature engineering, model training, backtesting, risk management, and live order routing. In addition, our book integrates probability, Bayesian logic, and statistical learning foundations throughout.

Machine Trading: Deploying Computer Algorithms to Conquer the Markets (Chan 2017)

The differences are substantial in orientation, scope, and modernity. Our book reveals a striking expansion in methodological breadth and depth. While *Machine Trading* covers traditional regression and ARIMA techniques, our book advances these models through modern regularization methods such as Ridge, Lasso, and Elastic Net. Bayesian inference—absent in Chan's text—is introduced comprehensively using packages like MCMCpack, rstan, and brms, allowing probabilistic modeling and uncertainty quantification. Deep learning, treated only superficially in *Machine Trading*, is fully developed in our book through keras and torch, extending applications to time series, sentiment analysis, and nonlinear feature extraction. Reinforcement learning, entirely missing from *Machine Trading*, is implemented in our book via PPO agents in R, reflecting cutting-edge adaptive trading research. Similarly, clustering and unsupervised learning are expanded beyond minimal discussion to include k-means, Gaussian Mixture Models (GMM), and Rmixmod, enabling the discovery of latent structures in financial data. Finally, our book introduces interpretability and explainable AI frameworks—lime, DALEX, and iml—which are crucial for transparency in financial modeling but absent from Chan's work. In essence, our book transforms the scope of algorithmic trading education from classical econometrics to a comprehensive AI/ML-driven paradigm.

How this book is structured

Algorithmic Trading and AI/ML with R follows a clearly defined, pedagogical structure that builds from foundational R tools to full automation and integration with APIs.

Chapter 1 introduces the essential R packages that form the backbone of the machine learning and algorithmic trading ecosystem in R. It begins by covering general-purpose machine learning frameworks such as caret and mlr, which offer comprehensive tools for model training, tuning, and evaluation across a variety of algorithms. For more advanced modeling, the chapter highlights deep learning libraries like keras and torch, which provide interfaces for building neural networks and leveraging GPU acceleration. Bayesian methods are also discussed, with packages like rstan and MCMCpack enabling users to perform probabilistic modeling and inference using Markov Chain Monte Carlo techniques. Explainability tools are introduced to help interpret and understand complex model predictions, an important aspect for transparency in trading strategies. The chapter then shifts focus to algorithmic trading and backtesting, reviewing popular R packages such as quantmod, TTR, and PortfolioAnalytics that facilitate data retrieval, technical indicator calculation, and portfolio risk management. Finally, a comparison between Python and R for algorithmic trading is provided, emphasizing R's strength in statistical modeling and financial analysis, while noting Python's broader ecosystem and versatility in production and system integration. This overview establishes a foundation for using R effectively in developing, testing, and deploying quantitative trading strategies based on machine learning.

Chapter 2 focuses on the acquisition, structuring, and management of financial market data essential for quantitative analysis and trading. It explains how to obtain data using various APIs such as Yahoo Finance for free market quotes, FRED for economic time series, Interactive Brokers (IBKR) for comprehensive real-time and historical data, and Schwab for brokerage-specific information. The chapter details key R packages that simplify data retrieval and manipulation, including **quantmod** for financial modeling, tidyquant for integrating tidyverse principles with financial data, alphavantager for accessing Alpha Vantage's API, and Rblpapi for Bloomberg data extraction. Instructions on handling both real-time streaming data and historical datasets are provided, emphasizing best practices for organizing large volumes of data, managing time series formats, and ensuring data integrity. This foundation enables robust market data workflows crucial for building reliable trading models and backtesting strategies.

Chapter 3 addresses the essential techniques for data wrangling and efficient storage in the context of financial analytics. It covers the use of different data structures in R such as data `frames`, `tibbles`, and the highly optimized `data.table`, discussing their respective advantages for handling large datasets and performing fast data manipulation. Attention is given to integrating with both local and remote data storage solutions, focusing on modern data lakes and file formats that ensure scalability and performance. The chapter highlights tools like `DuckDB` for local SQL-based querying on large datasets, and efficient binary formats such as Feather and fst that enable rapid reading and writing of data. Emphasizing best practices in organizing, cleaning, and storing data, this chapter prepares readers to manage complex financial datasets effectively, ensuring smooth downstream analyses and model development.

Chapter 4 centers on exploratory data analysis and visualization techniques specifically tailored for trading signals. It introduces key R packages such as ggplot2 for creating versatile and customizable static graphics, plotly for interactive visualizations that enhance data exploration, and the PerformanceAnalytics R package for calculating and visualizing financial performance metrics. The chapter also covers methods for recognizing candlestick patterns, which are vital for technical analysis, and demonstrates techniques for identifying volatility clustering—periods where price volatility groups together, often signaling market regime changes. By combining these tools and methods, this chapter equips readers to uncover meaningful patterns in financial data, validate trading hypotheses, and communicate insights effectively through compelling visual representations.

Chapter 5 introduces the fundamental concepts of machine learning as applied to trading, focusing on both supervised and unsupervised learning techniques. It explores regression and classification methods using powerful R packages such as glmnet for regularized linear models, randomForest for ensemble-based decision trees, and gradient boosting frameworks like xgboost and lightgbm that deliver high predictive accuracy. The chapter also covers unsupervised learning techniques like principal component analysis (PCA) and clustering methods, which are essential for detecting market regimes and uncovering latent structures in financial data. These tools provide a solid foundation for building predictive models and developing adaptive trading strategies that respond dynamically to changing market conditions.

Chapter 6 explores reinforcement learning as a powerful approach for developing adaptive trading strategies through agent-based methods. It outlines the reinforcement learning workflow in R, demonstrating how trading agents can be trained to interact dynamically with market environments by learning policies that maximize cumulative rewards. The chapter emphasizes reward optimization techniques tailored to financial objectives and discusses handling the complexities of dynamic, non-stationary market conditions. By applying reinforcement learning principles, traders can build models that continuously learn and adapt to evolving price patterns and market regimes, enabling more responsive and robust trading systems.

Chapter 7 provides a comprehensive guide to integrating with the Interactive Brokers (IBKR) API for live trading directly from R. It covers key functions such as **`twsConnect`** for establishing a connection, **`reqMktData`** and **`reqRealTimeBars`** for streaming market data, and **`placeOrder`** for submitting live orders. The chapter includes practical examples demonstrating live order placement, order logging to track executions and statuses, and robust error handling techniques to manage communication issues and API exceptions. A complete mapping of IBKR API functions tailored for R users is presented, empowering traders to automate order management, monitor real-time market conditions, and implement dynamic trading strategies within a seamless brokerage workflow.

Chapter 8 builds upon earlier material to focus on the automation and monitoring of algorithmic trading systems. It explores the design of resilient strategy loops capable of continuously executing trades while adapting to changing market conditions. The chapter emphasizes the importance of comprehensive monitoring and logging frameworks to track system performance, record trade activity, and detect anomalies in real time. It also discusses fail-safe automation techniques to ensure stable operation and quick recovery from errors or connectivity issues. Best practices for robustness and maintaining continuous operation are presented, highlighting methods for error handling, process supervision, and automated alerts to support reliable, long-term deployment of algorithmic trading strategies.

Chapter 9 serves as a companion to the *QuantRoom* YouTube channel (`@quantroom`), providing video tutorials and scripts that complement the book's content. It features detailed walkthroughs on integrating with Interactive Brokers and Schwab brokerage platforms, demonstrating practical steps for live trading system setup and management. The chapter bridges theoretical concepts with real-world implementation by offering educational content that guides users through strategy development, coding, and deployment. This resource enhances learning by combining visual explanations with hands-on examples, supporting a deeper understanding of algorithmic trading workflows and techniques.

Appendix A—C++ and R Integration presents a detailed technical comparison between the **IBrokers R API** and the **TWS C++ API**, focusing on their functional mappings such as requests, callbacks, and order handling. It analyzes the performance trade-offs associated with each, particularly regarding multi-threading capabilities and latency considerations. The section also provides practical guidance on using **Rcpp** to seamlessly integrate C++ code into R, enabling low-latency execution critical for high-frequency and production-grade trading systems. This appendix offers developers a clear blueprint to migrate prototypes developed in R into robust, high-performance production environments leveraging C++ efficiencies.

Appendix B—R C++ Application Programming Interface (API). The purpose of this appendix is to demonstrate, through concrete and fully reproducible examples, how modern **Rcpp** integrates C++ code within the R environment. While the body of this book emphasizes algorithmic design, statistical modeling, and performance optimization, the ability to compile and link C++ functions directly into R represents the essential bridge between conceptual modeling and high-performance implementation. By showing how a simple source file can be transformed into a dynamically loaded library via `Rcpp::sourceCpp()`, the appendix illustrates R's full extension pipeline—from preprocessing and compilation through automatic registration and memory-safe exchange of data objects. These examples reinforce the principles of reproducibility, portability, and modularity that underlie professional-grade quantitative finance research, ensuring that readers can both prototype rapidly and deploy computationally intensive routines using the same unified R/C++ workflow.

Cover

Rīga, the capital of Latvia, is famous for its exceptional Art Nouveau architecture, which flourished from 1900 to 1914. This movement was driven by the city's rapid economic growth and its status as a major port and industrial center. Architects, many of whom studied at the Rīga Polytechnic Institute, embraced the new aesthetic that spread across Europe.

While influenced by trends like German Jugendstil and Viennese Secession, Rīga developed distinct variations, including Eclectic, Perpendicular, National Romantic, and Neo-Classical styles. The buildings are characterized by elaborate façades adorned with intricate floral motifs, mythological figures, geometric patterns, and vibrant colors, transforming streets into open-air museums.

The left-panel shows an Art Nouveau façade featuring a stylized maskaron with a serene expression, framed by swirling hair-like patterns. The right-hand panel (top) shows a sculpture from a building corner depicting a knight in armor holding a shield. The right-hand panel (middle) shows an ornate column featuring a maskaron hidden within a mass of carved leaves and garlands. The right-hand panel (bottom) shows a portion of the Freedom Monument (Brīvības piemineklis), which blends elements of National Romanticism and Modernism, featuring sculptural groups that represent Latvian history.

Supplemental Website and YouTube Channel

All the book's scripts can be accessed on the book's GitHub branch at `https://github.com/jgvRa/Rbook2`. The *QuantRoom* YouTube channel

(`@quantroom`) provides video tutorials and scripts that complement the book's content. This method showcases real-time problem-solving, and delivers a more engaging and accessible way to master algorithmic trading using R and the Schwab Trader API.

Author biographies

Jason Guevara is a financial analyst and accountant. He maintains a YouTube channel (`https://www.youtube.com/@quantroom`) dedicated to developing practical R scripts to assist active traders and R quants. Jason also does contract work for OIS Market Research Group as an R financial systems architect, coder, and developer. Jason provides a unique blend of financial expertise and coding experience to the quant finance field. Jason holds a Bachelor of Science degree in Finance and a minor in Economics from California State University (CSU)–Northridge (2014). Jason's passion for markets began during the Great Recession. The rise of algorithmic trading at that time ignited his passion which to date continues to fuel his productivity. Jason uses his R programming skills to craft algorithmic trading scripts for personal exploration, research, and applications. He has been programming in R since 2012. Jason's dedicated YouTube channel is the premier guide for traders looking to master R in finance. By sharing his expertise online, he equips traders with the confidence to navigate the complex field of algorithmic trading.

Ričards Bulavs is a graduate with a Bakalaurs finansēs [B.Sc. in Finance] from The University of Latvia (2025). Ričards joined the OIS Market Research Group in July 2025 as a Research Associate 'Analyzing Financial Market Data, Implementing Financial Models, and trading equity, index, and futures instruments using quantitative methods and machine trading.' Prior to joining the OIS Market Research Group, Ričards was a student at the Emerio & Lourdes Linares Research and Education Center, where he learned to use the Interactive Brokers (IBKR) Trader Workstation (TWS) for trading equity, index, and futures instruments. Ričards successfully mastered TWS and the TWS API. He also mastered minimal-model (MinMod) trading tactics and option strike price selection using the Greeks and stochastic differential equation (SDE) derived empirical probability distributions. Born in Jūrmala, a Latvian resort city on the Gulf of Rīga, Ričards provides a unique blend of financial knowledge and quant coding experience in C++, Python, and R. Ričards specializes in crafting algorithmic trading strategies for exploration, research, and applications.

Dr. Oskars Linares is the Founder (2015), Research Director, and Quant Strategist at the OIS Market Research Group, an investment collective specializing in premium generation across equity, index, and futures options. A member of the Econometric Society and International Institute of Forecasters, Dr. Linares developed the proprietary Minimal-Model (MinMod) to guide OIS trading operations. His technical toolkit includes an SDE ARIMA-variant forecaster, which leverages empirical probability distributions and Bayesian updating to optimize strike price selection.

Oskars began his mathematical modeling career at the University of Michigan, Ann Arbor under the dual mastery of Professors Jeffrey B. Halter and John A. Jacquez. Training under Jacquez—a pioneer in compartmental analysis—Dr. Linares focused on nonlinear differential equations modeling and began a 30-year tenure with R (migrating from S-PLUS in 1995). His quest for mathematical rigor led him to the National Institutes of Health (NIH) in Bethesda, where he worked with Dr. Loren A. Zech at the Laboratory of Mathematical Biology. There, he utilized SAAM (Simulation, Analysis, And Modeling) and S-PLUS to navigate biomathematical complexity. This journey culminated at UPENN with Dr. Raymond C. Boston, applying Bayesian multilevel models to repeated measurement data to manage stochastic instability. With over 100 peer-reviewed scientific abstracts presented at Society meetings and papers to his name, Dr. Linares has ensured his models have remained statistically robust.

Oskars has also published several book chapters, and is co-author of the first editions of Investigating Biological Systems Using Modeling (Academic Press, 1999), Plain English for Doctors and Other Medical Scientists (Oxford University Press, 2017), Diagnosing and Treating Medicus Incomprehensibilis (Oxford University Press, 2018), Prescriptions for Quant Traders Using R: Videos and Scripts (CRC Press, 2026). He received the Great Seal of the United States Award (1993) for his advancements in mathematical-medicine research on aging. Oskars now lives in Rīga, Latvija.

Acknowledgments

No book is written in isolation, and this one is no exception. We are especially grateful to our editor Lara Speiker for not only her useful comments on earlier drafts of this book, but also for her encouragement and extreme patience in seeing this work reach its conclusion.

We are deeply grateful to our production editor, Rachael Panthier, for her expert guidance and steadfast support throughout the development of this work. Her insightful advice on the technical layout and the overall structure of the manuscript was instrumental in refining the presentation. Beyond her editorial acumen, Rachael's patience and attention to detail ensured that the complex integration of our work was translated into a cohesive and polished final product. We owe much of the clarity and professional finish of these chapters to her dedicated oversight.

We extend our sincere gratitude to the reviewers whose insightful comments and constructive critiques greatly improved the clarity and quality of this work.

First and foremost, I (JG) express my deepest gratitude to God for granting me the opportunity to share this work and for placing extraordinary and talented individuals in my path who have been instrumental in bringing this book to its final form.

To Jamahl Johnson of Alpha Beta Financial Technologies Inc., thank you for entrusting me with your projects and inspiring me to grow as an R programmer. Your keen eye for detail and innovative trading style have profoundly influenced this work, and I am grateful for your encouragement and inspiration.

To my subscribers at `https://www.youtube.com/@quantroom`, thank you to all 8.5K of you for being part of the *QuantRoom* community! Your support means the world to me—from the thoughtful comments and great questions to the way you've helped grow this channel into a real hub for data-driven investing, quant research, and intelligent market discussions. Whether you've been here since the first video or joined recently, I appreciate every view, like, and share. I'm excited to continue exploring the world of quantitative finance with you all—this is just the beginning!

To my Patreon supporters, I extend my sincere appreciation for your steadfast support over the past few years. Your thoughtful questions and messages have been a constant source of motivation, driving me to continue researching and creating tutorials to support your journeys. To all R programmers in finance, whether new or seasoned, I hope this book inspires you to push boundaries. I look forward to the day when I can learn from your insights as well. As the saying goes, a rising tide lifts all boats, and together, we navigate the vast sea of knowledge.

Jason Guevara
Santa Barbara, California
(April 11, 2026)

I (RB) would like to extend my sincere appreciation to the OIS Market Research Group for the opportunity to contribute to its ongoing research in quantitative finance and algorithmic trading. Joining the Group in July 2025 has been an intellectually rewarding experience—one that has deepened my understanding of financial modeling, machine trading, and data-driven decision-making.

I am especially grateful to Dr. Oskars Linares for his mentorship and guidance throughout my research journey. His emphasis on rigor, clarity, and innovation has shaped the way I approach both theoretical and applied aspects of quantitative finance. My time at the Emerio & Lourdes Linares Research and Education Center was equally formative. It was there that I first learned the foundational skills that later supported my work with minimal-model (MinMod) trading tactics and option strike price selection using stochastic differential equation (SDE) based probability models.

Born and raised in Jūrmala, I have always drawn inspiration from the rhythm and precision of nature—a perspective that now informs how I think about financial systems and algorithmic design. I remain grateful for the chance to blend finance and coding, applying C++, Python, and R to the art and science of market exploration.

Ričards Bulavs
Rīga, Latvia
(April 11, 2026)

First and foremost, my heartfelt thanks go to my wife (Annemarija Linaresa, MD/JD). Her unwavering encouragement, patience, and love have been my anchor throughout this process—this book is as much yours as it is mine. I (OL) acknowledges the use of Claude Code 2.0.27 (Sonnet 4.5) for coding, debugging, and documentation assistance throughout the development of this work.

I am deeply grateful to David Daly Jr., software engineer, formerly at Google and now with a hedge fund in Manhattan, New York, for his many engaging discussions on quantitative and algorithmic trading in finance, and for introducing me to Claude, which has since become an invaluable tool in my research and development work. I would also like to thank my students. Your curiosity, enthusiasm, and thoughtful questions have continually inspired me and challenged me to grow as an educator. The energy and insight you bring to our discussions have enriched my own understanding and make teaching all the more rewarding. Thank you for reminding me of the joy of learning and discovery. And last, but not least, to our readers—thank you for picking up this book and allowing our words to become a part of your world. Your curiosity and openness are the reason our work matters.

Dr. Oskar Linares
Rīga, Latvia
(April 11, 2026)

Typographical Conventions

The following typographical conventions are used in this book:

Italic
New terms and emphasized text.

`Constant width`
File names, file extensions, program elements such as arguments, variables, pointers, data types, environment variables, statements, and keywords, as well as database names, database tables, email addresses, and URLs.

`Constant width bold`
R functions, methods, and commands.

Bold font
R packages.

Sans Serif
Code listings, ticker symbols, class names, and objects.

List of Figures

List of Tables

Listings

1

Key AI/Machine Learning R Packages

You should not aspire to enter the worlds of others, but build your own and try to lure in these rare, truly exceptional and special characters.

(Petr Aven)

1.1 Introduction

Artificial Intelligence (AI) and Machine Learning (ML) are dramatically transforming financial markets by significantly enhancing predictive modeling, risk assessment, and decision-making processes. Alongside these groundbreaking technologies, algorithmic trading has become indispensable in modern finance, enabling the development, backtesting, and execution of systematic strategies at an unprecedented scale. This chapter brings together essential R packages that empower AI/ML workflows and bolster algorithmic trading infrastructure, presenting them in a cohesive narrative complemented by a summary table for easy reference.

1.2 Algorithmic Trading and AI/ML Packages

1.2.1 General ML Frameworks

caret, mlr3, tidymodels

These three ecosystems represent the backbone of machine learning in R, providing standardized workflows for model development, evaluation, and deployment. **caret** (Classification and Regression Training) was one of the earliest attempts to unify training and resampling under a consistent interface. Building on this foundation, **mlr3** offers a modern, object-oriented framework with support for parallelization and advanced tuning strategies. Meanwhile, **tidymodels** leverages the tidyverse philosophy, emphasizing composability and clean syntax for data preprocessing, model fitting, and validation. Together, they enable reproducible pipelines that abstract away low-level details and let practitioners focus on model design and interpretation.

randomForest, ranger

Ensemble learning using decision trees is a cornerstone of applied machine learning, and **randomForest** is the classic implementation in R. It aggregates many decision trees to reduce variance and improve generalization. **ranger** serves as a faster, more memory-efficient reimplementation, optimized for high-dimensional and large-scale datasets. These packages are particularly effective for tabular data where nonlinear interactions and variable importance measures are critical, making them widely used in financial and scientific applications.

xgboost, lightgbm

Gradient boosting frameworks have become the workhorses of structured data modeling. **xgboost** pioneered the modern boosting paradigm in R, offering scalable training and strong predictive performance. **lightgbm** improves training speed and memory efficiency through histogram-based algorithms and leaf-wise tree growth, making it highly suitable for large datasets.

glmnet

Regularization is essential for controlling model complexity, especially in high-dimensional financial datasets. **glmnet** implements efficient algorithms for *LASSO*, *Ridge*, and *Elastic Net regression*, allowing practitioners to shrink coefficients, perform variable selection, and balance bias-variance tradeoffs. Its speed and stability make it an indispensable tool when working with large predictor spaces or collinear variables, as often encountered in econometrics, portfolio modeling, and text-based financial signals.

e1071

The **e1071** package is a versatile toolkit that provides implementations of classical machine learning algorithms such as *support vector machines* (SVMs), *Naive Bayes* classifiers, and `clustering methods` like *k-means*. Its SVM module, in particular, is a robust choice for classification tasks where decision boundaries are nonlinear and margin maximization is desired. For practitioners, it serves as a gateway to statistical learning methods that predate the deep learning era but remain competitive in structured data applications.

nnet

For neural network modeling in R, **nnet** offers a straightforward interface to single-hidden-layer feedforward networks and multinomial logistic regression. Although relatively simple compared to modern deep learning frameworks, it is useful for smaller problems where interpretability and computational efficiency are valued. In financial contexts, it is often applied to classification tasks such as regime detection or categorical forecasting, providing a lightweight neural modeling option within the broader R ecosystem.

1.2.2 Deep Learning

keras, tensorflow

The **keras** and **tensorflow** packages provide a high-level and low-level interface, respectively, for deep learning in R. **keras** acts as a user-friendly wrapper that streamlines model definition, training, and evaluation, while **tensorflow** exposes the underlying computational graph for greater flexibility and scalability. Together, they enable practitioners to build complex architectures such as *convolutional neural networks* (CNNs), *recurrent neural networks* (RNNs), and *transformer-based models*. In finance, these tools are increasingly applied to sequential modeling of time series, natural language processing of financial texts, and nonlinear feature extraction.

torch

The **torch** package brings native **PyTorch** functionality into R, offering an efficient tensor library with GPU support. Unlike **keras**, **torch** integrates tightly with R's functional programming style, making it appealing to researchers who prefer low-level control over model training. It allows for rapid prototyping of custom deep learning architectures while still benefiting from automatic differentiation and parallelized computation. Torch is particularly well-suited for experimental modeling where flexibility and custom loss functions are required.

h2o

h2o provides a distributed and scalable machine learning platform accessible from R. Its deep learning module supports multilayer feedforward neural networks with adaptive tuning, while the broader ecosystem covers *gradient boosting*, *generalized linear models*, and *AutoML*. The *AutoML* functionality is especially valuable, as it automates model selection and hyperparameter optimization, enabling practitioners to quickly identify strong baselines. For financial applications, **h2o** is often used in production pipelines where model scalability and deployment readiness are priorities.

1.2.3 Bayesian Methods

rstan, brms

rstan offers direct access to the Stan probabilistic programming language, providing tools for specifying and sampling from Bayesian hierarchical models. **brms** builds on top of **rstan**, giving users a formula-based interface similar to **lme4**, which lowers the barrier to entry for Bayesian modeling. These packages make it possible to quantify uncertainty rigorously, incorporate prior knowledge, and perform posterior predictive checks. In finance, they are commonly applied to risk modeling, hierarchical time-series forecasting, and robust portfolio inference.

MCMCpack

MCMCpack supplies a suite of Bayesian regression and generalized linear models through pre-coded Markov Chain Monte Carlo (MCMC) routines. It is particularly useful for practitioners who want Bayesian inference without writing full Stan code. While less flexible than **rstan** or **brms**, it remains effective for standard Bayesian models and quick exploratory analysis. Its role in financial machine learning is often in constructing predictive distributions for asset returns or macroeconomic variables.

1.2.4 Explainability

lime, DALEX, iml

The rise of complex models has made interpretability a key requirement in financial applications where accountability is critical. The **lime** package adapts the *Local Interpretable Model-agnostic Explanations framework* to R, enabling users to approximate and visualize local decision boundaries. **DALEX** (*Descriptive mAchine Learning EXplanations*) extends interpretability to global model behavior, offering diagnostic plots, feature importance, and breakdown profiles. **iml** (*Interpretable Machine Learning*) provides a unified interface for explainability methods, including Shapley values, partial dependence plots, and permutation feature importance. Collectively, these tools allow practitioners to validate, audit, and communicate model behavior an essential step in regulated industries like finance.

1.2.5 Algorithmic Trading Packages

quantmod

The **quantmod** package is a cornerstone of quantitative finance in R, designed for rapid prototyping and modeling of financial time series. It provides tools to download, chart, and manipulate market data, and includes functions for building technical indicators. For algorithmic traders, **quantmod** often serves as the entry point into the R ecosystem, supporting both exploratory analysis and data preparation for more advanced modeling.

tidyquant

tidyquant bridges the gap between **tidyverse** workflows and financial data analysis. It wraps quantmod's market data retrieval in a tidy framework, enabling seamless integration with `dplyr`, `ggplot2`, and other **tidyverse** packages. This makes it particularly effective for building reproducible data pipelines, time-series visualizations, and strategy backtests that align with modern data science practices.

IBrokers

IBrokers provides an R interface to the Interactive Broker's API, enabling programmatic access to real-time and historical market data as well as automated trade execution. By linking directly to a brokerage platform, it turns R into a live trading environment. This package is essential for practitioners moving from research to deployment, where strategy execution and order routing must be tested under real market conditions.

alphavantager

The **alphavantager** package integrates R with the Alpha Vantage API, a popular provider of free and low-cost financial data. It supports equities, forex, and cryptocurrency data, making it a flexible choice for multi-asset strategy research. While not as fast as premium data feeds, **alphavantager** is ideal for prototyping and academic projects where cost and accessibility are primary concerns.

Rblpapi

Rblpapi connects R to the Bloomberg API, granting access to institutional-grade financial data, including equities, fixed income, derivatives, and economic indicators. Its strength lies in high-quality, real-time feeds and deep historical coverage. For professional trading operations, **Rblpapi** supports integration with Bloomberg terminals and is crucial for sophisticated portfolio management and risk modeling.

1.2.6 Strategy and Backtesting

Backtest

backtest is a modern replacement for **quantstrat**, providing a flexible framework for specifying, simulating, and evaluating trading strategies in R. It supports event-driven workflows, risk controls, and custom performance metrics. Its modular design makes it suitable for both academic exploration and applied trading research, while maintaining compatibility with tidyverse-style data pipelines.

PortfolioAnalytics

PortfolioAnalytics has evolved into the primary package for portfolio analysis and optimization in R. It provides functionality for strategy evaluation, constraint handling, and portfolio risk assessment, effectively absorbing many of the accounting roles previously handled by blotter. Its integration with **PortfolioAnalytics** enables detailed evaluation of strategy-level and portfolio-level outcomes.

Custom Instrument Management

Instead of relying on the retired **FinancialInstrument** package, modern workflows often manage instrument metadata directly within **tidyverse** structures such as **tibble** or **data.table**. Combined with packages like **quantmod**, **tidyquant**, and **IBrokers**, this approach ensures flexible integration with diverse data sources while maintaining clarity and reproducibility in multi-asset strategies.

strand

The **strand** package is a modern R framework for portfolio and strategy backtesting that emphasizes speed, transparency, and integration with **tidyverse** workflows. Unlike legacy packages such as **quantstrat** and **blotter**, which have become difficult to maintain, **strand** offers a lightweight and extensible design. It allows users to define strategies as tidy data frames of positions, transactions, and cash flows, which are then processed to simulate portfolio evolution. This approach reduces the complexity of rule-based scripting and instead provides a clear, data-centric model of trading activity.

Key features of **strand** include:

- Vectorized backtesting: Efficient handling of time-series trades without heavy looping.
- Tidy integration: Strategies and results are expressed as tibbles, making them compatible with **dplyr**, **tidyr**, and **ggplot2**.
- Custom metrics: Easy calculation of returns, drawdowns, Sharpe ratios, and other performance measures.
- Flexibility: Users can design strategies across multiple assets, apply custom position sizing rules, and extend the framework with additional analytics.

For algorithmic trading research, **strand** is particularly valuable as a modern alternative to the deprecated **quantstrat/blotter** stack. It provides a balance between simplicity for prototyping and robustness for production-grade strategy evaluation.

1.2.7 Risk and Performance

PerformanceAnalytics

PerformanceAnalytics is one of the cornerstone R packages for financial analysis. It provides an extensive suite of tools for evaluating portfolio performance, risk-adjusted returns, and drawdowns. The package includes functions for computing metrics such as *Sharpe ratio*, *Sortino ratio*, *maximum drawdown*, *Value-at-Risk* (VaR), and *Conditional VaR* (CVaR), along with visualization tools for return distributions, cumulative returns, and risk-return scatter plots. Its design integrates seamlessly with **xts** and **zoo** objects, making it particularly suited for time-series financial data. In practice, **PerformanceAnalytics** remains the *de facto* standard in R for benchmarking strategies, producing professional-grade performance reports, and supporting robust risk management workflows.

FRAPO, riskmetric, cvar, fPortfolio

This collection of packages covers specialized risk modeling and optimization. **FRAPO** provides functions for financial risk modeling and portfolio optimization, while **riskmetric** implements methodologies for volatility and covariance estimation. The **cvar** package focuses on Value-at-Risk models, a standard risk metric in finance, and **fPortfolio** offers advanced portfolio optimization and asset allocation methods. Together, they extend the capabilities of R for institutional-grade risk management.

rugarch, rmgarch

rugarch and **rmgarch** specialize in univariate and multivariate GARCH models, respectively. These packages allow for *volatility forecasting*, *correlation modeling*, and *scenario analysis*, which are essential for derivative pricing, risk assessment, and stress testing. They are widely used in both academic research and professional trading to model time-varying volatility in asset returns.

1.2.8 Execution and Integration

RQuantLib

RQuantLib provides access to the QuantLib C++ library, a leading framework for quantitative finance. It supports option pricing, fixed income analytics, and risk management models. For algorithmic trading, **RQuantLib** brings sophisticated pricing engines into R, enabling accurate valuation and hedging of complex instruments.

plumber

plumber turns R scripts into REST APIs with minimal effort, making it possible to deploy trading strategies and analytics as web services. This capability is critical for integrating R-based research into production systems, where strategies must interface with external platforms or monitoring dashboards in real time.

shiny

shiny is a framework for building interactive web applications directly from R. It enables the creation of live dashboards for monitoring strategy performance, visualizing market data, or managing risk in real time. For practitioners, **shiny** bridges the gap between research outputs and user-facing tools, supporting both transparency and operational control.

In the context of algorithmic trading, the tools tabulated in Table 1.1 are not merely for prediction. Whether you are deploying gradient-boosted ensembles or performing Bayesian inference for risk management, the R ecosystem offers a comprehensive suite of frameworks that bridge the gap between academic research and production-grade execution. Table 1.1 summarizes the key packages for AI/ML, performance attribution, and the deployment of systematic strategies.

TABLE 1.1: AI/ML and algorithmic trading packages in R

Category	**Package(s)**	**Notes / Typical Use**
General ML Frameworks	caret, mlr3, tidymodels	Unified ML workflows, tuning, resampling
Ensembles	randomForest, ranger, xgboost, lightgbm, catboost	Tree-based and gradient boosting models
Regression	glmnet, e1071, nnet	Regularized regression, SVM, neural nets
Deep Learning	keras, tensorflow, torch, h2o	DL with TensorFlow, PyTorch, AutoML
Bayesian ML	rstan, brms, MCMCpack	Bayesian modeling and inference
Explainability	lime, DALEX, iml	AI tools for model interpretation
Market Data	quantmod, tidyquant, IBrokers, alphavantager, Rblpapi	Financial market data APIs
Strategy & Backtesting	PortfolioAnalytics	Backtesting, accounting, portfolio analysis
Risk & Performance	PerformanceAnalytics, FRAPO, RiskMetrics, VaR, fPortfolio, rugarch, rmgarch	Risk analysis, VaR, volatility models
Execution	RQuantLib, plumber, shiny	Pricing, deployment, dashboards

1.2.9 Conclusion

The above section highlighted key R packages used in Artificial Intelligence, Machine Learning, and Algorithmic Trading. These tools offer a complete workflow, from acquiring market data and developing machine learning models to backtesting, risk assessment, and deployment. Table 1.1 serves as a quick reference guide for practitioners who are building intelligent financial systems in R.

1.3 Modern Comparative Analysis of Python vs. R for Algorithmic Trading

1.3.1 Introduction

Algorithmic trading requires programming languages that effectively balance research capabilities, performance, and live execution. While Python has become the industry standard, R has evolved into a fully competitive alternative. With modern libraries such as **tidyverse**, **tidyquant**, **Backtest**, **PortfolioAnalytics** and **PerformanceAnalytics** replacing older frameworks, R now offers traders a robust ecosystem for data analysis, strategy design, and portfolio optimization that stands alongside Python in both research and applied settings.

1.3.2 Python Ecosystem and Libraries

Core Libraries

numpy provides numerical computation, enabling matrix operations and linear algebra at scale. **pandas** is the standard for tabular and time-series data manipulation, widely used in trading. **scipy** extends numerical routines with optimization, statistics, and signal processing. **statsmodels** supports econometrics and statistical modeling such as regressions and ARIMA.

Trading and Backtesting

backtrader is a widely used backtesting engine that supports live trading, portfolio management, and integration with broker APIs. **zipline** (originally developed by Quantopian) provides a framework for event-driven backtesting. **vectorbt** enables fast, vectorized backtests directly on pandas DataFrames. **quantstats** and **pyfolio** generate comprehensive performance and risk reports for trading strategies.

Machine Learning and AI

scikit-learn is the workhorse ML library for regression, classification, and clustering. **tensorflow** and **pytorch** provide deep learning support with GPU acceleration, essential for reinforcement learning and deep trading models.

Visualization

matplotlib underpins most Python visualization workflows. **seaborn** builds on **matplotlib**, providing higher-level statistical plots. **plotly** allows interactive, browser-based visualization. **bokeh** enables web-based interactive dashboards.

APIs and Broker Connectivity

ib_insync simplifies Interactive Brokers API integration. **alpaca-trade-api** provides access to commission-free stock trading. **ccxt** integrates with hundreds of cryptocurrency exchanges. **oandapy** connects to the OANDA FX and CFD trading platform.

1.3.3 R Ecosystem and Libraries (Modern Workflow)

Core Libraries

data.table provides ultra-fast data manipulation, ideal for tick and intraday data. **dplyr** is the tidyverse tool for relational data manipulation. **xts** and **zoo** define extensible time-series objects that underlie most R finance packages.

Trading and Backtesting

tidyquant bridges tidyverse data workflows with quantitative finance, simplifying financial data analysis. **quantmod** handles financial data retrieval and basic charting, often the first step in strategy prototyping. **Backtest** is a newer framework for strategy testing, easier to learn than legacy quantstrat. **PortfolioAnalytics** enables portfolio-based backtesting, optimization, and factor exposures.

Risk and Performance Analytics

PerformanceAnalytics provides a comprehensive suite for risk metrics, return analysis, and visualization of performance. **TTR** offers technical indicators such as moving averages, RSI, MACD, and Bollinger Bands. **RQuantLib** connects R to the QuantLib C++ library for options pricing, bond analytics, and risk measures.

Machine Learning and AI

tidymodels is the modern ML framework in R, replacing caret with tidyverse alignment. **mlr3** supports extensive model pipelines and hyperparameter tuning. **keras** and **tensorflow** R interfaces provide access to deep learning frameworks.

Visualization

ggplot2 is the standard for statistical and time-series visualization in R. **plotly** extends ggplot2 with interactivity. **lattice** remains useful for multivariate visualization and conditioning plots.

Broker APIs

IBrokers is the R interface for Interactive Broker's API, enabling market data and order routing (see Table 1.2).

TABLE 1.2: Comparison of python and modern R ecosystems for algorithmic trading

Domain	**Python**	**R (Modern)**
Ecosystem	Broad: ML, APIs, trading	Modernized: tidyquant, PortfolioAnalytics
Backtesting	`backtrader`, `zipline`, `vectorbt`	`Backtest`, `tidyquant`, `PortfolioAnalytics`
Performance	Parallel + GPU scaling	Strong time series, simulation focus
Community	Large, industry-dominant	Academic + quant finance niche
APIs	Broad broker/exchange support	Limited, mainly IB via `IBrokers`
Visualization	`matplotlib`, `seaborn`, `plotly`	`ggplot2`, `plotly`, `lattice`
Risk Analysis	Moderate	Excellent (`PerformanceAnalytics`, `PortfolioAnalytics`)
Machine Learning	Excellent (`scikit-learn`, DL frameworks)	Strong (`tidymodels`, `mlr3`, `keras`)

1.3.4 Conclusion and Recommendation

Python continues to be a leading choice for production-ready algorithmic trading due to its strong integration with broker APIs, extensive machine learning frameworks, and scalable infrastructure. However, R is quickly gaining ground, thanks to modern packages like **tidyquant**, **Backtest**, and **PortfolioAnalytics**, which have made it a competitive option for strategy research, portfolio analytics, and systematic backtesting. The two ecosystems are now more similar than ever, with the decision between them increasingly based on team expertise and project goals rather than strict technical limitations.

For professional traders, a hybrid development approach may be most effective: R serves as a platform for strategy research, statistical modeling, and data exploration; Python offers scalability, API integration, and execution efficiency; and, as discussed in the Appendix, C++ may provide a performance-critical layer required for low-latency computation and real-time trading infrastructure.

2

Market Data Acquisition

> I don't dream at night, I dream all day; I dream for a living.
>
> *(Steven Spielberg)*

2.1 Introduction

Market data acquisition is the foundation of algorithmic trading workflows. In R, a diverse set of packages provides traders with access to equities, derivatives, macroeconomic, and cryptocurrency data. These packages connect to both free and institutional-grade sources, offering a bridge between raw market feeds and strategy development. This chapter reviews the most important R packages for handling market data in algorithmic trading, outlining their capabilities and practical applications.

2.2 Core Market Data Packages

2.2.1 quantmod

The **quantmod** package is the cornerstone of financial data analysis in R. It simplifies downloading equity and macroeconomic data from sources such as Yahoo Finance and FRED, provides convenient charting functions, and includes tools for technical indicators. **quantmod** is often the first step in building reproducible data pipelines for trading strategies.

2.2.2 tidyquant

tidyquant integrates **quantmod**'s functionality with the **tidyverse** ecosystem. It allows users to acquire market data in tidy formats, making it easy to manipulate and visualize using **dplyr** and **ggplot2**. This package is particularly useful for building systematic, pipeline-driven workflows in algorithmic trading. At its core, the function `tq_get()` allows seamless access to multiple data sources. These include historical stock prices, exchange traded funds, indices, and mutual funds from Yahoo Finance, macroeconomic indicators from the Federal Reserve Economic Data (FRED), foreign exchange rates, and certain fundamental data (via backends such as **quantmod**). By returning all results in tidy data frames (tibble objects), **tidyquant** simplifies the process of importing, wrangling, and analyzing diverse financial time series in a manner consistent with modern R data science practices.

2.2.3 IBrokers

IBrokers connects R to the Interactive Brokers API. It provides access to historical and real-time market data and supports automated order execution. For practitioners, **IBrokers** is essential when transitioning from research to live trading, as it enables strategies to be tested directly against broker feeds.

2.2.4 Charles Schwab (Trader) API

Charles Schwab offers an API that provides traders with access to real-time quotes, historical market data, account information, and order management. As of the latest public disclosures, The Charles Schwab Corporation (ticker SCHW) reports approximately 36.9 million active brokerage accounts as of February 28, 2025. Although this API is not as well-established in the R ecosystem as Interactive Brokers, it can still be accessed through RESTful endpoints. This enables integration with R using packages such as **httr** or **curl** for making HTTP requests, and **jsonlite** for parsing the responses. As a result, users can create custom functions to stream market data or download historical prices directly into their R workflows.

The *Schwab (Trader) API* is particularly appealing to U.S. retail traders because it offers commission-free trading for equities and ETFs, including stocks, ETFs, options, and mutual funds. While there isn't a dedicated R wrapper like **IBrokers**, its REST-based architecture ensures compatibility with R's versatile data-handling tools. This allows traders to incorporate Schwab's data feeds into backtesting, portfolio monitoring, and automated trading research.

We (JG) provide detailed videos and scripts on *Getting Started with the Charles Schwab (Trader) API & R* (`https://www.youtube.com/watch?v=A0iFYj5iM5U&t=12s`), and its use for algorithmic trading. Additionally, we (JG) created a Shiny App to assist active traders usage of the *Schwab (Trader) API* (`https://www.youtube.com/watch?v=bUR8RO6MMEM&t=3s`).

Comparison: Interactive Brokers vs. Charles Schwab

Both Interactive Brokers and Charles Schwab offer APIs that provide access to market data and order management. However, they differ in terms of scope, maturity, and integration with R. *Interactive Brokers* provides a direct R interface through the **IBrokers** package, supporting equities, futures, options, and forex. This makes it well-suited for multi-asset and institutional-grade strategies. Its ecosystem is mature, allowing for seamless integration into backtesting and execution workflows, although it requires careful setup of Trader Workstation or IB Gateway.

In contrast, *Charles Schwab's API* is REST-based and primarily focused on U.S. equities, ETFs, and options. While it does not have a dedicated R wrapper, its compatibility with general HTTP and JSON packages makes it accessible to developers familiar with building custom connectors. Schwab is particularly appealing for retail traders because of its commission-free equity trades, while Interactive Brokers excels in global coverage, professional-grade features, and extensive instrument support.

Ultimately, Interactive Brokers is the stronger choice for advanced, multi-asset algorithmic trading in R, while Schwab offers a simpler and more cost-effective option for equity-focused strategies (see Table 2.1).

TABLE 2.1: Comparison of Broker APIs for Market Data in R

Feature	**Interactive Brokers (IB)**	**Charles Schwab**
R Integration	Dedicated **IBrokers** package provides direct access to market data and order execution.	No dedicated R package; accessed via REST API using `httr`, `curl`, and `jsonlite`.
Asset	Equities, ETFs, options, futures, forex, bonds; strong global market coverage.	Equities, ETFs, options, mutual funds; primarily U.S. market focused.
Data Access	Real-time and historical data feeds via TWS or IB Gateway.	Real-time quotes, historical data, and account info via REST endpoints.
Execution	Full order management (market, limit, stop, conditional orders).	Supports order entry and account-level integration for equities and options.
Strengths	Mature ecosystem, institutional-grade features, broad multi-asset support.	Commission-free equities/ETFs, accessible REST interface, cost-effective for retail traders.
Limitations	Requires setup of Trader Workstation (TWS) or Gateway; learning curve for API.	Limited asset classes, no dedicated R wrapper, more suited for U.S. equities only.

Continued on next page

Feature	Interactive Brokers (IB)	Charles Schwab
Best Fit	Professional/quant traders, multi-asset strategies, global markets.	U.S.-based retail traders, equity/ETF strategies, simpler automation needs.

2.2.5 Rblpapi

Rblpapi is an interface to the Bloomberg API, widely used by institutional investors. It provides access to real-time and historical market data, as well as macroeconomic and alternative datasets. **Rblpapi**'s integration ensures low-latency data feeds for professional-level trading and portfolio management.

2.2.6 alphavantager

alphavantager links R to the Alpha Vantage API, offering equities, forex, and cryptocurrency data. While not as comprehensive as Bloomberg, its free and low-cost tiers make it ideal for prototyping, academic research, and entry-level algorithmic trading projects.

2.2.7 Quandl

Quandl provides a unified interface to a broad range of traditional and alternative financial datasets. From equities and derivatives to macroeconomic series, **Quandl** consolidates multiple vendors under one API. Premium subscriptions unlock access to institutional-quality datasets.

2.2.8 crypto2, cryptowatchR

For cryptocurrency data, packages like **crypto2** and **cryptowatchR** provide access to prices, volumes, and order book data across multiple exchanges. They are critical for researchers and traders working in the digital asset space.

2.2.9 xts and zoo

xts and **zoo** define the time-series structures on which most financial R packages are built. They support flexible indexing, merging, and manipulation of irregularly spaced data, making them indispensable for structuring raw market feeds.

2.2.10 data.table

data.table is not finance-specific, but its speed and efficiency make it ideal for handling large tick or intraday datasets. In algorithmic trading, it is frequently used to preprocess and aggregate high-frequency data.

To effectively implement an algorithmic trading workflow, selecting the appropriate data acquisition layer is critical for ensuring both low-latency execution and reproducible analysis. Table 2.2 outlines the foundational R ecosystem for market data, ranging from general-purpose visualization tools to specialized high-frequency and cryptocurrency interfaces.

TABLE 2.2: Key R packages for market data acquisition in algorithmic trading

Category	Package(s)	Notes / Typical Use
General Market Data	quantmod, tidyquant	Acquisition, visualization, tidyverse integration
Broker APIs	IBrokers, Rblpapi, alphavantager	Interactive Brokers, Bloomberg, Alpha Vantage interfaces
Specialized Sources	Quandl, crypto2, cryptowatchR	Macro, derivatives, cryptocurrency markets
Time Series Structures	xts, zoo, data.table	Core structures for financial time series and intraday data

2.2.11 Conclusion

Market data packages form the backbone of algorithmic trading in R. From free APIs such as Yahoo finance and Alpha Vantage to institutional feeds like Bloomberg, these tools allow traders to construct robust, reproducible workflows. Combined with powerful data structures from **xts**, **zoo**, and **data.table**, they ensure that even high-frequency or multi-asset strategies can be efficiently supported within R's ecosystem.

2.3 Data Storage Solutions

2.3.1 Introduction

Efficient data storage is a critical component of algorithmic trading infrastructure, since market data can range from daily bars to terabytes of tick-level history. R users have multiple approaches, depending on scale and use case:

2.3.2 SQLite and PostgreSQL

For structured time series or tabular data, relational databases such as `SQLite` and `PostgreSQL` are reliable solutions. `SQLite` is lightweight, server-less, and easy to embed within R workflows, making it suitable for prototyping and personal research. `PostgreSQL`, on the other hand, scales to enterprise settings, supports advanced indexing, and integrates well with extensions like `timescaledb` for high-frequency time-series workloads. Both can be accessed in R via packages such as **RSQLite** and **RPostgres**.

2.3.3 Parquet, Feather, and FST

Columnar storage formats like `Parquet`, `Feather`, and `FST` are increasingly popular for high-performance data pipelines. They provide fast read/write operations, compression, and efficient use of memory. In R, the **arrow** package allows seamless use of `Parquet` and `Feather`, while `FST` offers blazing-fast binary storage for time-series datasets. These formats are particularly useful for intraday or tick data, where speed of retrieval matters more than relational joins.

2.3.4 Cloud Storage and Data Lakes

Cloud-based solutions, such as AWS S3, Google Cloud Storage, or Azure Blob, provide scalable, cost-effective ways to store large datasets while enabling remote collaboration. Through R packages like **aws.s3**, **googleCloudStorageR**, or **AzureStor**, market data can be streamed directly from the cloud into backtesting pipelines. This model is increasingly attractive for distributed research groups or when strategies require petabyte-scale data.

2.3.5 Hybrid Approach

In practice, many trading teams adopt a hybrid architecture: lightweight relational databases (`SQLite`/`Postgres`) for structured daily data, fast columnar formats (`Parquet`/`FST`) for high-frequency streams, and cloud storage for long-term archiving. This layered approach balances accessibility, speed, and scalability, ensuring that historical research and live trading both remain efficient.

2.3.6 MongoDB

`MongoDB` is a document-oriented NoSQL database that stores data in flexible JSON-like structures. For algorithmic trading, this makes it especially well-suited for heterogeneous or nested market data such as order books, tick-level event logs, or trade metadata. Its ability to scale horizontally across clusters supports very large datasets, and its schema flexibility accommodates evolving strategy needs without costly schema migrations. In R, MongoDB can be accessed through the mongolite package, allowing efficient queries and aggregation pipelines directly from within trading workflows. The trade-off is that complex joins and relational-style queries are less efficient than in SQL databases, but for unstructured or semi-structured financial data, MongoDB is a powerful solution.

2.3.7 DuckDB

DuckDB is an embedded analytical database often described as the "SQLite for analytics." Unlike traditional relational databases optimized for transactions, DuckDB is optimized for analytical queries on large datasets, with performance comparable to columnar stores such as Parquet. It integrates seamlessly with R via the **duckdb** package, enabling fast SQL queries on in-memory or on-disk data without a server process. For algorithmic trading, DuckDB excels in exploratory research, backtesting on large time-series datasets, and integrating with modern file formats like Parquet. Its single-file architecture and SQL interface make it both lightweight and powerful, particularly for researchers who need a middle ground between SQLite's simplicity and PostgreSQL's scalability (see Table 2.3).

Once your data is acquired via the R packages previously discussed, the next critical step is selecting a storage architecture that balances retrieval speed with structural integrity. Whether you are managing high-frequency tick data or conducting exploratory research, the choice of database dictates the efficiency with which you can access large datasets. Table 2.3 evaluates various storage solutions—ranging from lightweight embedded databases like DuckDB and SQLite to enterprise-grade relational and NoSQL systems—tailored for the unique I/O demands of algorithmic trading.

TABLE 2.3: Key data storage options for market data in algorithmic trading

System	Type	Strengths	Typical Use Case
SQLite	Relational (lightweight)	Serverless, simple, portable, easy R integration (RSQLite)	Small-scale research projects, daily OHLC data, personal prototyping
PostgreSQL	Relational (enterprise)	Scalable, robust indexing, supports timescaledb, integrates with RPostgres	Enterprise-level storage of intraday/long-term financial data, production systems

Continued on next page

System	Type	Strengths	Typical Use Case
MongoDB	NoSQL (document)	Schema flexibility, good for JSON-like tick-/order book data, R access via `mongolite`	Real-time event data, trade logs, unstructured or semi-structured market feeds
DuckDB	Analytical SQL (embedded)	Optimized for analytics, fast on Parquet/CSV, lightweight, R access via `duckdb`	Exploratory research, backtesting on large historical datasets, in-memory analytics
Parquet / FST	Columnar file formats	High compression, fast I/O, R access via `arrow` or `FST`	High-frequency time-series data, efficient read/write pipelines, archival storage
Cloud (S3, GCS, Azure)	Object storage / data lake	Scalable, distributed	Long-term archival, collaborative research, big-data pipelines across teams

2.3.8 Conclusion and Recommendations

Selecting the right data storage solution is critical for building efficient, scalable, and reliable algorithmic trading workflows. No single technology fits all use cases; rather, the choice depends on data volume, frequency, structure, and the stage of the trading pipeline.

For lightweight research or daily bar data, `SQLite` provides simplicity and portability, while `DuckDB` offers a powerful analytical option for larger historical datasets without requiring a server. For institutional environments or production pipelines, `PostgreSQL`, especially with the `timescaledb` extension, remains the most versatile choice for structured time-series data, delivering both scalability and reliability.

When dealing with unstructured or semi-structured datasets, such as order books or tick-level event logs, `MongoDB` stands out due to its schema flexibility and ability to handle complex, nested data formats. For high-frequency or compressed archival, columnar file formats like `Parquet` or `FST` should be the default option, ensuring fast I/O and efficient storage.

Finally, for long-term storage, team collaboration, and scaling beyond on-premise infrastructure, *cloud-based object storage* (e.g., AWS S3, Google Cloud Storage, or Azure) is recommended. These solutions integrate well with R and modern data pipelines, offering durability, elasticity, and global accessibility.

Recommendation: practitioners may adopt a hybrid architecture. Use `SQLite` or `DuckDB` for exploratory work, `PostgreSQL` for production time-series databases, `Parquet`/`FST` for high-frequency pipelines, `MongoDB` for event-driven datasets, and cloud storage for archiving and collaboration. This layered approach ensures that algorithmic trading systems remain flexible, performant, and scalable across both research and production environments.

2.4 Data Wrangling Packages for Algorithmic Trading

Before market data can be analyzed, modeled, or used in backtesting, it may need to be cleaned, aligned, and reshaped. Data wrangling is therefore a critical stage of algorithmic trading. In R, a number of packages support the efficient manipulation of financial time series and large-scale panel data, bridging the gap between raw acquisition and strategy-ready datasets.

2.4.1 Core Time-Series Structures

xts

The **xts** package (eXtensible Time Series) defines the primary time-series object used by nearly all financial packages in R. It provides a robust framework for indexing, subsetting, and merging irregular time-series data such as price quotes or returns. Because xts underpins packages like quantmod and PerformanceAnalytics, it is essential for algorithmic trading workflows.

zoo

zoo is the foundational package for irregular time series in R, offering flexible indexing and time-based operations. **xts** extends **zoo**, but **zoo** itself remains useful for lightweight manipulation and for integrating with base R time-series methods.

2.4.2 High-Performance Wrangling

data.table

The **data.table** package is designed for fast, memory-efficient handling of large datasets. In algorithmic trading, it is particularly valuable for intraday or tick-level data where joins, aggregations, and rolling operations must be performed at scale. Its syntax is concise, and its speed makes it the default choice for high-frequency financial research.

dplyr / tidyverse

The **dplyr** package, as part of the **tidyverse**, provides a declarative grammar of data manipulation (`filter`, `mutate`, `summarise`, `group_by`). Combined with **ggplot2** and other **tidyverse** tools, **dplyr** enables reproducible pipelines for preparing financial data for modeling or visualization.

2.4.3 Tidy Financial Wrangling

tidyquant

tidyquant integrates **tidyverse** syntax with financial time-series workflows, making it possible to apply **dplyr** verbs directly to `xts` objects. It includes functions like `tq_transmute` and `tq_mutate` for rolling calculations, technical indicators, and returns analysis. This package bridges traditional finance libraries (**quantmod**, **PerformanceAnalytics**) with modern **tidyverse** practices.

timetk

timetk extends **tidyverse** functionality with specialized time-series tools, enabling summarization, rolling window calculations, and visualization. It is particularly effective when preparing features for machine learning models or when conducting exploratory time-series analysis.

tsibble

tsibble provides a tidy framework for managing temporal data with explicit indexing and keys, making it ideal for panel data with multiple assets. It integrates with the **fable** ecosystem for forecasting and supports advanced handling of temporal hierarchies.

2.4.4 Date and Text Utilities

lubridate

The **lubridate** package simplifies parsing and manipulating dates and times, a common requirement in financial research where timestamps must be aligned across different markets and time zones.

janitor

janitor provides convenient functions for cleaning column names and tabular data, especially useful when importing messy financial datasets from APIs or CSVs.

stringr

The **stringr** package standardizes string manipulation tasks, such as parsing ticker symbols, identifiers, or other text-based financial fields.

Beyond acquisition and storage, the technical success of algorithmic trading hinges on the efficiency of the data transformation layer. In algorithmic trading, raw market feeds are rarely ready for immediate analysis; they require precise alignment, cleaning, and windowing. Whether you are executing high-frequency aggregations via `data.table` or performing exploratory tidy-style analysis, the R ecosystem provides a specialized suite of tools for these tasks. Table 2.4 categorizes the essential packages for wrangling financial data, ensuring that your pipeline remains both performant and reproducible.

TABLE 2.4: Key R packages for wrangling financial data

Category	**Package(s)**	**Notes / Typical Use**
Core Structures	xts, zoo	Time-series indexing, merging, and irregular frequency handling
High-Performance	data.table	Fast joins, aggregations, rolling ops; essential for tick-/intraday data
Tidy Wrangling	dplyr, tidyverse	Declarative manipulation, pipelines, integration with ggplot2
Finance-Tidy Bridge	tidyquant	Apply tidyverse syntax to xts; rolling indicators, returns
Time-Series Extension	timetk, tsibble	Panel data, summarization, ML feature prep, forecasting integration
Date Handling	lubridate	Parsing and aligning timestamps across markets
Cleaning Utilities	janitor, stringr	Cleaning column names, parsing tickers, handling text fields

2.5 Conclusion

Data wrangling is the bridge between raw acquisition and strategy development. For algorithmic traders, R offers a layered ecosystem: **xts** and **zoo** for time-series foundations, **data.table** and **dplyr** for efficient manipulation, and **tidyquant**, **timetk**, and **tsibble** for modern, tidyverse-compatible workflows. Utility packages like **lubridate**, **janitor**, and **stringr** streamline the cleaning of dates, names, and identifiers. Together, these tools ensure that financial datasets can be transformed from noisy, irregular inputs into reliable structures ready for modeling, backtesting, and execution.

3

Trading Models and Strategy Design

Yes: I am a dreamer. For a dreamer is one who can only find his way by moonlight, and his punishment is that he sees the dawn before the rest of the world.

(Oscar Wilde)

3.1 Trend Following

Trend following is based on the idea that securities with consistent price momentum tend to continue in the same direction (Figure 3.1-1). If a stock or index is rising steadily, it is more likely to keep rising rather than reverse, and the same applies to falling prices. This suggests a market inefficiency since, in a perfectly efficient market, price changes would be random. The challenge for traders is distinguishing between genuine trends and short-lived noise, while also being ready for periods of poor performance and frequent small losses, known as whipsaws.

3.1.1 Moving Averages and Crossovers

The moving-average crossover method helps identify trends by smoothing price data over specific periods. A short-term moving average reacts quickly to recent changes, while a long-term moving average shows the broader market direction, allowing traders to spot potential turning points.

A *Golden Cross* happens when the short-term moving average crosses above the long-term average, signaling a bullish trend. Conversely, a *Death Cross* occurs when it crosses below, indicating a bearish trend. These signals help traders systematize their decisions for entering and exiting positions.

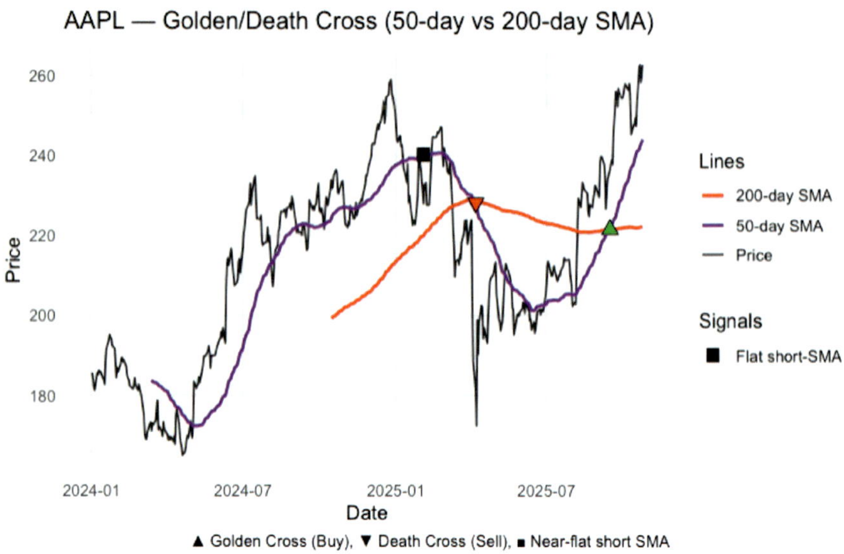

FIGURE 3.1-1: AAPL stock price (black line) over time with two important moving averages plotted: 50-day SMA (purple line) and the 200-day SMA (red line). Sell the inverted red triangle crossover and buy the green triangle crossover.

Figure 3.1-1 illustrates the occurrence of a *Death Cross*, where the short-term moving average falls below the long-term average. Periods of a near-flat short-term moving average are also highlighted, reflecting a transitional phase in market momentum.

3.1.2 Commentary

Trend-following strategies are effective partly due to predictable investor behaviors, notably herding, where investors follow the market direction instead of acting independently. This can lead to market prices deviating from their fundamental values in the short to medium term.

To implement trend-following, traders often use moving averages to gauge trends, which smooths out daily price fluctuations. Selecting the appropriate time frame is essential: short windows react quickly but can produce false signals, while long windows reduce noise but may miss early reversals.

There is no universally reliable moving-average length, as effectiveness varies across assets and market conditions. Consequently, traders must continuously experiment and adjust their strategies, combining an understanding of investor psychology with adaptability to changing market scenarios.

Listing 3.1.1: Death cross signal using short-term and long-term simple moving averages (SMAs).

```
# Required packages
# install.packages(c("tidyquant", "TTR", "dplyr", "ggplot2"))

library(tidyquant)
library(TTR)
library(dplyr)
library(ggplot2)

# Function to calculate SMA
calculate_sma <- function(data, window) {
  SMA(data, n = window)
}

# Function to calculate and plot Bollinger Bands
plot_bollinger_bands <- function(ticker, window, num_std, start_date,
     end_date) {

  # Download stock prices
  df <- tq_get(ticker, from = start_date, to = end_date) %>%
    select(date, adjusted) %>%
    rename(price = adjusted)

  # Calculate SMA and Std Dev
  df <- df %>%
    mutate(
      sma   = calculate_sma(price, window),
      sd    = runSD(price, n = window),
      upper = sma + num_std * sd,
      lower = sma - num_std * sd
    )

  # Generate buy/sell signals
  df <- df %>%
    mutate(
      buy_signal  = (price <= lower & lag(price) > lag(lower)),
      sell_signal = (price >= upper & lag(price) < lag(upper))
    )

  # Plot
  p <- ggplot(df, aes(x = date, y = price)) +
    geom_line(color = "black", size = 0.8, alpha = 0.8) +
    geom_line(aes(y = sma, color = "SMA"), size = 0.8) +
    geom_line(aes(y = upper, color = "Upper Band"), linetype = "dashed"
    ) +
    geom_line(aes(y = lower, color = "Lower Band"), linetype = "dashed"
    ) +
    geom_point(data = df %>% filter(buy_signal), aes(y = price),
               shape = 24, color = "green", fill = "green", size = 3) +
```

```
    geom_point(data = df %>% filter(sell_signal), aes(y = price),
               shape = 25, color = "red", fill = "red", size = 3) +
    labs(
      title = paste0(ticker, " Price with Bollinger Bands (", window, "
-day SMA, ", num_std, " Std Dev)"),
      x = "Date", y = "Price"
    ) +
    scale_color_manual(values = c("SMA" = "blue", "Upper Band" = "
darkred", "Lower Band" = "darkgreen")) +
    theme_minimal()

  print(p)
}

# Example run
plot_bollinger_bands("GOOG", 20, 2, "2024-10-01", "2025-12-31")
```

3.2 Mean Reversion

Mean reversion posits that asset prices tend to revert toward a long-run average, often described as "fair value." Deviations from this average may present trading opportunities, particularly in range-bound regimes.

3.2.1 Bollinger Bands and Thresholds

Traders can use moving averages and Bollinger Bands to estimate fair value and define statistically significant deviations. Parameters should be tuned cautiously to avoid overfitting.

Figure 3.2-1 highlights potential buy signals (green upward triangles) when prices fall below the lower band and potential sell signals (red downward triangles) when prices rise above the upper band, illustrating volatility-adjusted trading opportunities.

3.2.2 Commentary

Mean reversion strategies exploit market overreactions, leading to temporary price deviations from intrinsic values. They perform well in range-bound markets but struggle in trending ones. Defining "fair value" and detecting significant deviations are key challenges. Traders utilize tools like moving averages and Bollinger Bands for fair value proxies and statistical thresholds, requiring careful optimization and backtesting to avoid overfitting.

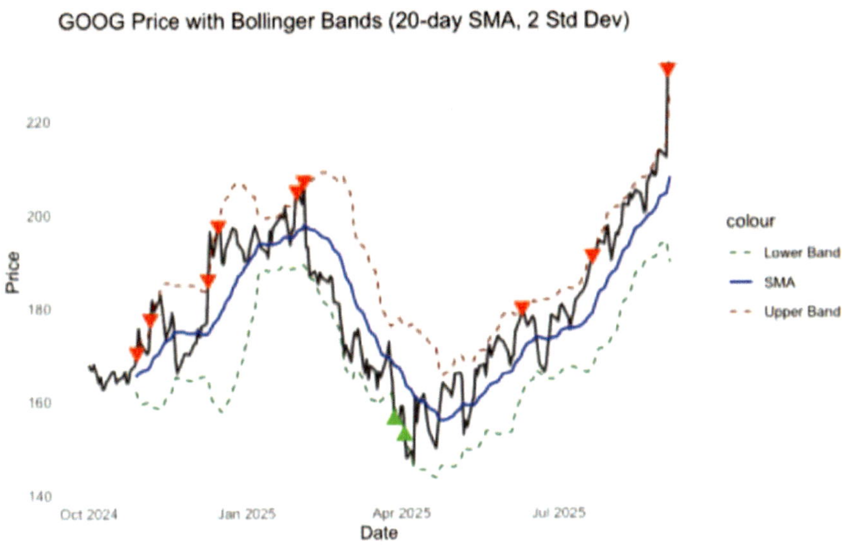

FIGURE 3.2-1: GOOG daily closing prices with 20-day simple moving average (SMA) and Bollinger Bands at ±2 standard deviations.

Listing 3.2.1: Bollinger bands at ±2 standard deviations.

```
# Required packages
# install.packages(c("tidyquant", "TTR", "dplyr", "ggplot2"))

library(tidyquant)
library(TTR)
library(dplyr)
library(ggplot2)

# Function to calculate SMA
calculate_sma <- function(data, window) {
  SMA(data, n = window)
}

# Function to calculate and plot Bollinger Bands
plot_bollinger_bands <- function(ticker, window, num_std, start_date,
    end_date) {

  # Download stock prices
  df <- tq_get(ticker, from = start_date, to = end_date) %>%
    select(date, adjusted) %>%
    rename(price = adjusted)

  # Calculate SMA and Std Dev
  df <- df %>%
```

```
    mutate(
      sma   = calculate_sma(price, window),
      sd    = runSD(price, n = window),
      upper = sma + num_std * sd,
      lower = sma - num_std * sd
    )

  # Generate buy/sell signals
  df <- df %>%
    mutate(
      buy_signal  = (price <= lower & lag(price) > lag(lower)),
      sell_signal = (price >= upper & lag(price) < lag(upper))
    )

  # Plot
  p <- ggplot(df, aes(x = date, y = price)) +
    geom_line(color = "black", size = 0.8, alpha = 0.8) +
    geom_line(aes(y = sma, color = "SMA"), size = 0.8) +
    geom_line(aes(y = upper, color = "Upper Band"), linetype = "dashed"
    ) +
    geom_line(aes(y = lower, color = "Lower Band"), linetype = "dashed"
    ) +
    geom_point(data = df %>% filter(buy_signal), aes(y = price),
               shape = 24, color = "green", fill = "green", size = 3) +
    geom_point(data = df %>% filter(sell_signal), aes(y = price),
               shape = 25, color = "red", fill = "red", size = 3) +
    labs(
      title = paste0(ticker, " Price with Bollinger Bands (", window, "
    -day SMA, ", num_std, " Std Dev)"),
      x = "Date", y = "Price"
    ) +
    scale_color_manual(values = c("SMA" = "blue", "Upper Band" = "
    darkred", "Lower Band" = "darkgreen")) +
    theme_minimal()

  print(p)
}

# Example run
plot_bollinger_bands("GOOG", 20, 2, "2024-10-01", "2025-12-31")
```

3.3 Statistical Arbitrage (Pairs Trading)

Statistical arbitrage strategies aim to exploit short-term price misalignments between assets that typically move together. One common method is pairs trading, where a trader identifies two co-moving assets and monitors their price spread (Figure 3.3-1). This spread usually stays around a historical average, but when it deviates beyond certain statistical limits, it signals a trading opportunity. The strategy involves selling the overpriced asset and buying the underpriced one, with the expectation that the spread will eventually revert to normal.

FIGURE 3.3-1: Illustration of a pairs trading strategy using Microsoft (MSFT) and Apple (AAPL).

In Figure 3.3-1, the black line shows the ratio of MSFT's price to AAPL's price over time, with the blue line representing the 20-day moving average. The dashed orange lines indicate two standard deviations above and below the mean. A sell signal (red triangle) is triggered when the ratio exceeds the upper band, suggesting a potential decline, while a buy signal (green triangle) occurs when it falls below the lower band, indicating a possible rise. This illustrates how traders use moving averages and standard deviation bands for entry and exit points in pairs trading.

3.3.1 Pairs Trading

Pairs trading is a market-neutral strategy that profits from the relative performance of two assets that historically move together. When their relationship diverges, traders buy the undervalued asset and sell the overvalued one, expecting a return to the mean. However, this relies on the assumption that the correlation between the assets remains stable. If that correlation weakens or breaks due to market shifts or news, the trade may not converge, leading to potential losses.

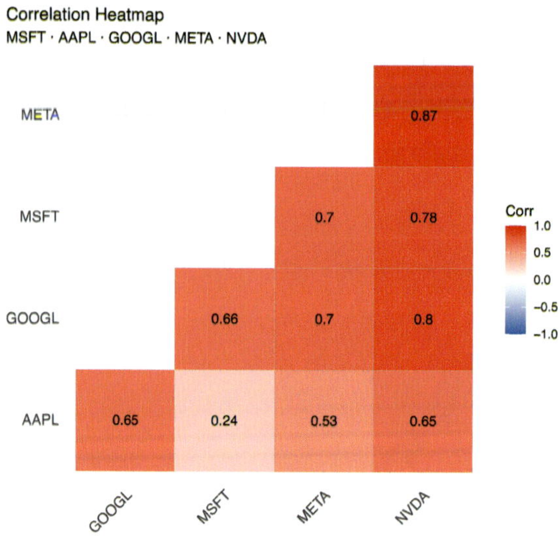

FIGURE 3.3-2: Numbered heat map of correlations among MSFT, AAPL, GOOGL, META, and NVDA from January 2024 to September 2025.

3.3.2 Commentary

A correlation matrix is valuable for pairs trading, as it helps identify asset relationships. The correlation heat map in Figure 3.3-2 shows the pairwise correlations among MSFT, AAPL, GOOGL, META, and NVDA from January 2024 to September 2025. The strongest correlation is between META and NVDA ($\rho = 0.87$), followed by GOOGL and NVDA ($\rho = 0.80$) and MSFT and NVDA ($\rho = 0.78$). The weakest correlation is between AAPL and MSFT ($\rho = 0.24$), indicating more independent price movements. However, a correlation matrix does not ensure a profitable pair.

Listing 3.3.1: Statistical arbitrage—pairs trading.

```
library(tidyquant)
library(dplyr)
library(tidyr)
library(zoo)
library(ggplot2)

download_data <- function(tickers, start_date, end_date) {
  tq_get(tickers, get = "stock.prices", from = start_date, to = end_
    date) %>%
    select(symbol, date, adjusted) %>%
    pivot_wider(names_from = symbol, values_from = adjusted) %>%
    arrange(date)
}
```

```
# ---- 2) Correlation matrix of price levels (pairwise complete) ----
calculate_correlation <- function(df_wide_prices) {
  #' Calculates correlation matrix of numeric columns (tickers).
  num <- df_wide_prices %>% select(-date)
  cor(num, use = "pairwise.complete.obs")
}

plot_pairs_trading <- function(ticker1, ticker2, window, num_std, start
    _date, end_date) {

  tickers <- c(ticker1, ticker2)
  prices  <- download_data(tickers, start_date, end_date)
  if (nrow(prices) == 0L || !all(tickers %in% colnames(prices))) {
    message("Could not retrieve data for one or both tickers.")
    return(invisible(NULL))
  }

  ratio <- prices %>%
    transmute(date, ratio = .data[[ticker1]] / .data[[ticker2]]) %>%
    mutate(
      mean  = zoo::rollapply(ratio, width = window, FUN = mean, align =
     "right", fill = NA_real_),
      sd    = zoo::rollapply(ratio, width = window, FUN = sd,   align =
     "right", fill = NA_real_),
      upper = mean + num_std * sd,
      lower = mean - num_std * sd,
      buy_signal  = ratio < lower,
      sell_signal = ratio > upper
    )

  # ---- dynamic legend labels (build FIRST) ----
  lab_ratio <- "Price Ratio"
  lab_mean  <- sprintf("%d-day Mean", window)
  lab_up    <- sprintf("+%g Std Dev", num_std)
  lab_dn    <- sprintf("-%g Std Dev", num_std)

  color_values <- setNames(
    c("black", "blue", "orange", "orange"),
    c(lab_ratio, lab_mean, lab_up, lab_dn)
  )

  color_breaks <- c(lab_ratio, lab_mean, lab_up, lab_dn)

  p <- ggplot(ratio, aes(x = date)) +
    geom_line(aes(y = ratio, color = lab_ratio), size = 0.9) +
    geom_line(aes(y = mean,  color = lab_mean),  size = 0.9) +
    geom_line(aes(y = upper, color = lab_up),    linetype = "dashed") +
    geom_line(aes(y = lower, color = lab_dn),    linetype = "dashed") +
```

````
    geom_point(
      data = subset(ratio, buy_signal),
      aes(y = ratio, shape = "Buy Signal"),
      size = 3, color = "green", fill = "green"
    ) +
    geom_point(
      data = subset(ratio, sell_signal),
      aes(y = ratio, shape = "Sell Signal"),
      size = 3, color = "red", fill = "red"
    ) +
    scale_color_manual(
      name   = "Lines",
      values = color_values,
      breaks = color_breaks
    ) +
    scale_shape_manual(
      name   = "Signals",
      values = c("Buy Signal" = 24, "Sell Signal" = 25),
      breaks = c("Buy Signal", "Sell Signal")
    ) +
    guides(
      color = guide_legend(nrow = 1, byrow = TRUE, order = 1),
      shape = guide_legend(nrow = 1, byrow = TRUE, order = 2)
    ) +
    labs(
      title = paste("Pairs Trading Strategy:", ticker1, "vs", ticker2),
      x = "Date", y = "Price Ratio"
    ) +
    theme_minimal(base_size = 13) +
    theme(
      legend.position     = "bottom",
      legend.box          = "horizontal",
      legend.box.spacing  = grid::unit(0.2, "lines"),
      legend.margin       = ggplot2::margin(2, 2, 2, 2),
      legend.key.size     = grid::unit(0.55, "cm"),
      legend.text         = element_text(size = 9),
      legend.title        = element_text(size = 10),
      plot.margin         = ggplot2::margin(10, 12, 20, 12)
    )

  print(p)
  invisible(p)
}

# ---- Save to PDF (legend fits) ----
pdf("pairs_trading_MSFT_AAPL.pdf", width = 9, height = 6)
plot_pairs_trading("MSFT", "AAPL", 20, 2, "2024-01-01", "2025-12-31")
```
````
````

`````
```{r}
pdf("pairs_trading_MSFT_AAPL.pdf", width = 10, height = 6)
p <- plot_pairs_trading("MSFT","AAPL",20,2,"2024-01-01","2025-12-31")
p + ggtitle("Pairs Trading Strategy: MSFT vs. AAPL")
dev.off()
```

```{r}
# If you want correlations across a basket:
px <- download_data(c("MSFT","AAPL","GOOGL","META", "NVDA"), "
    2024-01-01", "2025-12-31")
calculate_correlation(px)
```

```{r}
# install.packages("ggcorrplot")
library(dplyr)
library(ggcorrplot)

# 1) Get your prices and correlation
px <- download_data(c("MSFT","AAPL","GOOGL","META","NVDA"),
                    start_date = "2024-01-01", end_date = "2025-12-31")

cor_mat <- px %>% select(-date) %>% cor(use = "pairwise.complete.obs")
cor_mat <- round(cor_mat, 2)  # round so the labels are 2-decimal

# 2) Plot a numbered, clustered, lower-triangle heatmap
p_corr <- ggcorrplot(
  cor_mat,
  lab = TRUE,                  # overlay numbers
  hc.order = TRUE,             # hierarchical clustering for nicer
    ordering
  type = "lower",              # lower triangle
  method = "square",           # square tiles
  colors = c("#4575B4", "#F7F7F7", "#D73027"),  # blue-white-red
  outline.color = "grey80",
  ggtheme = ggplot2::theme_minimal(base_size = 12),
  show.legend = TRUE
) +
  ggplot2::labs(
    title = "Correlation Heatmap",
    subtitle = "MSFT AAPL GOOGL META NVDA",
    x = NULL, y = NULL
  ) +
  ggplot2::theme(
    plot.title.position = "plot",
    axis.text.x = ggplot2::element_text(angle = 45, hjust = 1),
    legend.position = "right"
  )
`````
`````

````
p_corr

# 3) Save to PDF
pdf("correlation_heatmap_FAANGish.pdf", width = 7, height = 6)
print(p_corr)
dev.off()

```

```{r}
pdf("3-1-3_sacorhm.pdf", width = 10, height = 6)
# install.packages("ggcorrplot")
library(dplyr)
library(ggcorrplot)

# 1) Get your prices and correlation
px <- download_data(c("MSFT","AAPL","GOOGL","META","NVDA"),
                    start_date = "2024-01-01", end_date = "2025-12-31")

cor_mat <- px %>% select(-date) %>% cor(use = "pairwise.complete.obs")
cor_mat <- round(cor_mat, 2)  # round so the labels are 2-decimal

# 2) Plot a numbered, clustered, lower-triangle heatmap
p_corr <- ggcorrplot(
  cor_mat,
  lab = TRUE,                 # overlay numbers
  hc.order = TRUE,            # hierarchical clustering for nicer
    ordering
  type = "lower",             # lower triangle
  method = "square",          # square tiles
  colors = c("#4575B4", "#F7F7F7", "#D73027"),  # blue-white-red
  outline.color = "grey80",
  ggtheme = ggplot2::theme_minimal(base_size = 12),
  show.legend = TRUE
) +
  ggplot2::labs(
    title = "Correlation Heatmap",
    subtitle = "MSFT AAPL GOOGL META NVDA",
    x = NULL, y = NULL
  ) +
  ggplot2::theme(
    plot.title.position = "plot",
    axis.text.x = ggplot2::element_text(angle = 45, hjust = 1),
    legend.position = "right"
  )

p_corr
````

```
# 3) Save to PDF
pdf("correlation_heatmap_FAANGish.pdf", width = 7, height = 6)
print(p_corr)
dev.off()
```

3.4 Cyclical

Stock prices often exhibit *cyclical behavior* unrelated to seasonal factors, driven by internal market forces like shifts in investor sentiment and reactions at key price levels. Recognizing and modeling these cycles can give traders an analytical edge.

3.4.1 Fast Fourier Transform

The Fast Fourier Transform (FFT) is a valuable tool for identifying cyclical-periodic type patterns. Originally used for signal processing, the FFT decomposes time series data, such as daily stock prices, into underlying frequency components. This allows traders to uncover repeating market rhythms that may not be apparent in raw price data.

TABLE 3.1: Dominant frequency components derived from FFT analysis of SPY's price cycles.

Freq.	Power	Period (days)
2	0.4458	2.2433
1	0.1124	8.8938
3	0.0587	17.0339

The FFT power spectrum of SPY's closing prices shown in Figure 3.4-1 reveals how variance is distributed across frequencies. Most of the spectrum is flat and noisy, but a few distinct peaks stand out. Table 3.1 summarizes the three dominant peaks, corresponding to periods of approximately 2.2 days, 8.9 days, and 17 days. Overall, the spectrum is primarily low-frequency and trend-like, but certain short- and medium-term cycles are consistently identifiable, as noted in Table 3.1.

3.4.2 Spectral Leakage Reduction

Figure 3.4-2 analyzes log-returns (detrended) and applies a *Hann window* to minimize spectral leakage, highlighting key peak periods in trading days. Raw price series are non-stationary due to long-term drift, so log-returns center the

(a) SPY closing price from 2020 to 2025, showing the post-pandemic recovery, subsequent corrections, and sustained upward trend into 2024–2025.

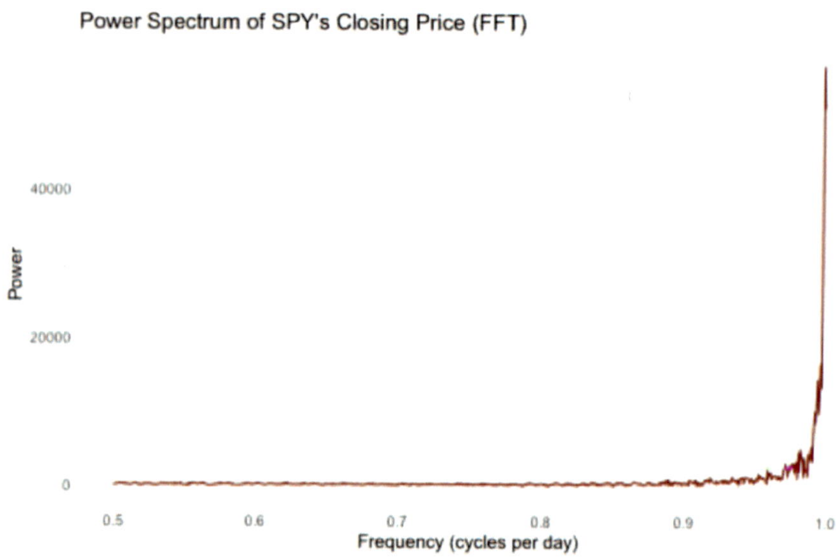

(b) Power spectrum of SPY's closing price series, obtained using the FFT.

FIGURE 3.4-1: SPY closing price from 2020 to 2025 alongside its frequency-domain representation using the FFT. The price chart highlights long-term upward trends with intermittent corrections, while the power spectrum shows that most variation occurs at low frequencies, consistent with trend-dominated market behavior.

series around zero, emphasizing short-term fluctuations. The Hann window smooths data edges to provide sharper spectral peaks. The spectrum is represented in frequency (cycles per day), but converted to trading periods for interpretation. For instance, a peak at 0.11 cycles/day corresponds to a 9-day cycle, indicating potential recurring market behavior. This process effectively turns raw price data into insights about cyclical patterns in trading behavior.

3.4.3 Commentary

The cyclical patterns identified by the FFT should not be mistaken for infallible forecasts of future prices. What they do provide, however, are useful clues about recurring rhythms in market behavior across particular historical intervals. A key point for students and practitioners alike is that the length of the time window chosen for the FFT matters greatly. Shorter windows may highlight faster, short-term cycles, while longer windows can uncover slower, more structural patterns. Because each window emphasizes a different aspect of market dynamics, analysts must align their choice of window with the trading horizon they care about.

Listing 3.4.1: Cyclical patterns identified by the Fast Fourier Transform (FFT).

```
# Required packages
# install.packages(c("quantmod", "ggplot2"))
library(quantmod)
library(ggplot2)

analyze_price_cycles <- function(ticker, start_date, end_date) {
  # Download stock data
  getSymbols(ticker, src = "yahoo", from = start_date, to = end_date,
    auto.assign = FALSE) -> data
  close_prices <- Cl(data)

  if (NROW(close_prices) == 0) {
    cat(sprintf("Could not retrieve data for %s.\n", ticker))
    return(NULL)
  }

  # FFT
  N <- NROW(close_prices)
  yf <- fft(as.numeric(close_prices))
  xf <- (0:(N - 1)) / N   # frequency in cycles per day (since daily
    data)

  # Prepare data for plots
  df_prices <- data.frame(Date = index(close_prices), Price = as.
    numeric(close_prices))
  df_fft <- data.frame(Freq = xf[(N %/% 2 + 1):N],
```

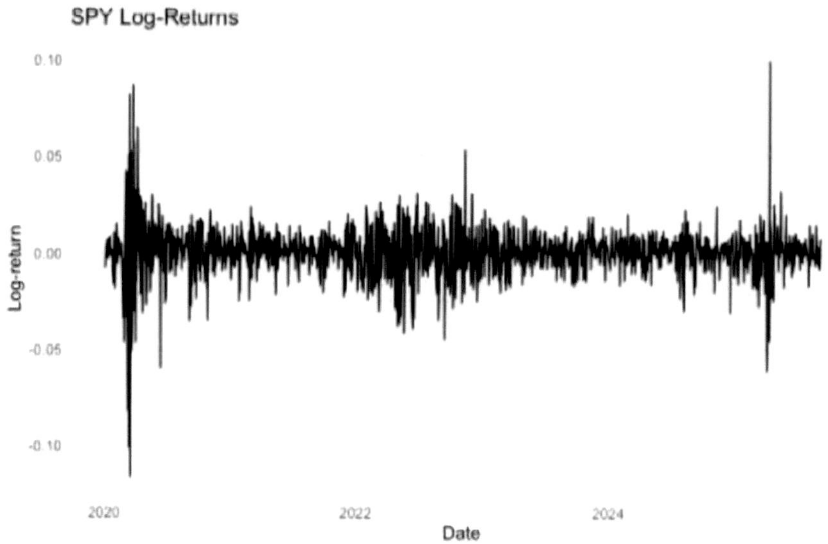

(a) SPY daily log-returns from 2020 to 2025, showing extreme volatility during the 2020 market shock, followed by more stable fluctuations in subsequent years with occasional spikes reflecting episodes of market stress.

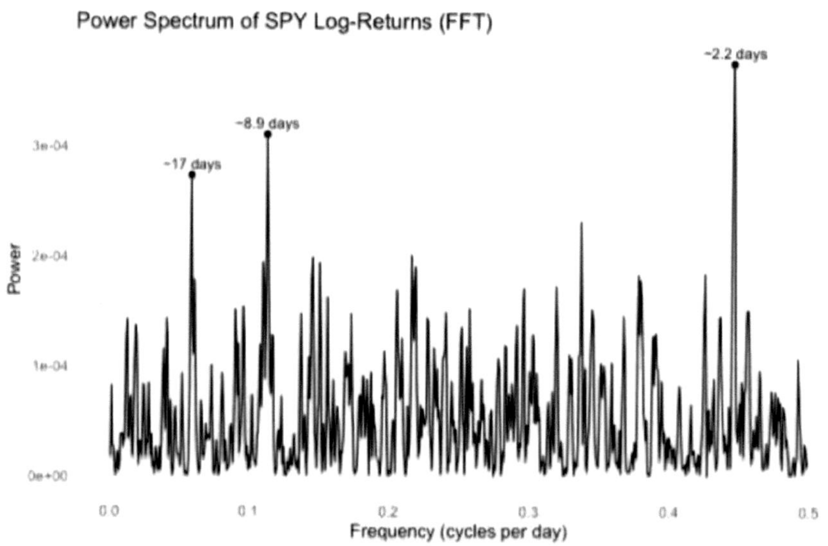

(b) Power spectrum of SPY log-returns (FFT). The spectrum reveals cyclical components at roughly 17-day, 8.9-day, and 2.2-day intervals, suggesting recurring short- to medium-term patterns in return volatility amid an otherwise noisy series.

FIGURE 3.4-2: SPY daily log-returns from 2020 to 2025 and their frequency-domain representation using the FFT.

```
                            Power = Mod(yf[(N %/% 2 + 1):N]))

  # Plot closing price
  p1 <- ggplot(df_prices, aes(x = Date, y = Price)) +
    geom_line(color = "steelblue") +
    labs(title = paste(ticker, "Closing Price"),
         x = "Date", y = "Price") +
    theme_minimal()

  # Plot FFT power spectrum
  p2 <- ggplot(df_fft, aes(x = Freq, y = Power)) +
    geom_line(color = "darkred") +
    labs(title = "Power Spectrum of SPY's Closing Price (FFT)",
         x = "Frequency (cycles per day)", y = "Power") +
    theme_minimal()

  # Display plots
  print(p1)
  print(p2)
}

# Example run
analyze_price_cycles("SPY", "2020-01-01", "2025-12-31")
```

3.5 Cluster

Stock and index levels often show a tendency to cluster around certain levels rather than being evenly distributed across the levels range. These clusters are meaningful because they frequently act as decision points for traders. When prices cluster at the lower end, they may create support—a level where buying interest is strong enough to slow or halt further declines. When prices cluster at the upper end, they may form resistance—a level where selling pressure tends to limit further advances.

Recognizing these clusters is valuable because they can signal potential inefficiencies in the market, where supply and demand temporarily misalign. Traders who can identify such levels may gain an advantage in anticipating price reversals or continuations.

3.5.1 Frequency Distribution Histogram

A practical way to uncover price clustering is to examine the historical frequency distribution histogram of a stock's trading prices. By plotting how often the stock has traded at different levels over a given period, one can see where trading activity has concentrated most heavily. These areas of high trading density often align with support or resistance zones, making them useful tools for market analysis.

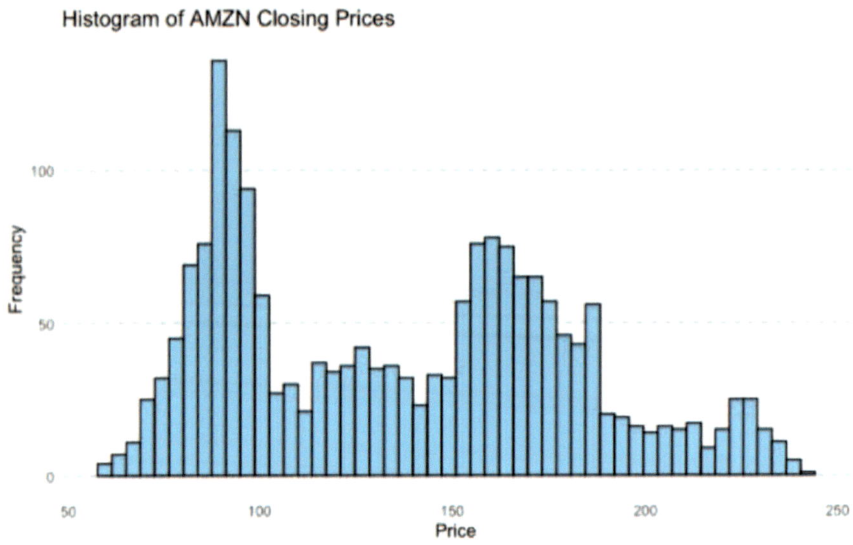

FIGURE 3.5-1: Frequency distribution histogram of Amazon (AMZN) daily closing prices.

Figure 3.5-1 displays the frequency distribution of AMZN highlighting prominent price clusters where the stock traded frequently, around \$95–105 and \$155–\$165, suggesting potential support and resistance zones shaped by repeated market activity.

3.5.2 Commentary

Traders often identify potential support and resistance levels visually, but their statistical strength and practical value are debatable. These levels tend to act as self-fulfilling prophecies, shaped by traders' collective expectations. The timeframe is crucial, as short samples reveal temporary levels while longer samples highlight significant, durable zones.

The histogram of Amazon's daily closing prices from 2018 to 2025 displayed in Figure 3.5-1 shows clustering around \$100 and \$160. Traders often see these clusters as potential support or resistance, reasoning that if prices gravitated there in the past, they may do so again. However, it's important to approach these interpretations with caution, as the predictive power of price clustering remains contested.

Listing 3.5.1: Histogram cluster.

```
library(tidyquant)
library(ggplot2)
library(dplyr)

identify_price_clusters <- function(ticker, start_date, end_date) {
  # Download stock price data
  data <- tryCatch({
    tq_get(ticker, get = "stock.prices", from = start_date, to = end_
     date)
  }, error = function(e) {
    message(sprintf("Could not retrieve data for %s", ticker))
    return(NULL)
  })

  if (is.null(data) || nrow(data) == 0) {
    message(sprintf("No data available for %s", ticker))
    return(invisible(NULL))
  }
  # Plot histogram of closing prices
  p <- ggplot(data, aes(x = close)) +
    geom_histogram(bins = 50, fill = "skyblue", colour = "black") +
    labs(
      title = paste("Histogram of", ticker, "Closing Prices"),
      x = "Price",
      y = "Frequency"
    ) +
    theme_minimal() +
    # Use a lighter grey instead of alpha (element_line has no alpha
     arg)
    theme(panel.grid.major.y = element_line(colour = "grey70", linetype
      = "dashed"))

  print(p)
  invisible(p)
}
# Example
identify_price_clusters("AMZN", "2018-01-01", "2025-12-31")
```

3.6 Chart Patterns

Chart patterns are recognizable visual formations on stock price charts that many traders believe offer insights into future price movements. Their potential significance, however, is contested. The key inefficiency lies in the possibility that these patterns may not reflect underlying economic forces but instead emerge as self-fulfilling prophecies. Because so many traders act on them, the collective behavior itself can reinforce the pattern and influence subsequent pricc action.

3.6.1 Double Top/Bottom

Strategies that rely heavily on chart patterns are often described as fragile. Their predictive power rests less on economic fundamentals than on trader psychology and the shared recognition of recurring shapes. Yet this very recognition can affect markets in meaningful ways. For instance (Figure 3.6-1), the *double top*, a bearish reversal formation, occurs when a stock makes two peaks near the same price level with a trough between them. A break below the trough is typically interpreted as a sell signal. Conversely, the *double bottom*, a bullish counterpart, consists of two troughs separated by a peak, where a break above the peak is taken as a signal to buy.

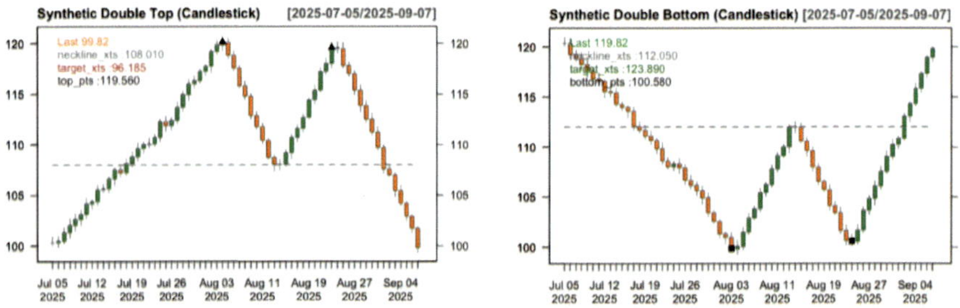

(A) Double top.

(B) Double bottom.

FIGURE 3.6-1: Comparison of two classical chart reversal patterns. The left panel illustrates a synthetic double top, where prices fail twice at the same resistance level before declining. The right panel shows a synthetic double bottom, where prices test support twice before reversing upward. Both patterns highlight how repeated failures at key levels can signal potential trend reversals.

3.6.2 Commentary

The reliability of chart patterns is often debated, as it varies by pattern type, analysis timeframe, and market conditions. A pattern that appears strong on a daily chart may seem less significant on shorter or longer timeframes. To enhance confidence in these signals, traders often use additional technical indicators like moving averages or momentum oscillators. Advanced applications even employ algorithms to quantify how closely price movements match ideal patterns, aiming to reduce subjectivity in chart analysis.

Listing 3.6.1: Double top and bottom chart patterns.

```
To minimize space. The complete version of the script is available for
    download on our website (https://github.com/jgvRa/Rbook2/chapter3/
    3-1-6_ChrtPatrn.Rmd).
```

3.7 Seasonality

The supply and demand dynamics of certain assets sometimes follow recurring, time-based patterns on daily, weekly, or monthly horizons. These seasonal effects can introduce short-term inefficiencies that attentive traders may seek to exploit. A well-known example is the tendency for asset managers to adjust their portfolios toward higher-quality holdings near the end of the month for reporting purposes.

3.7.1 Market Inefficiencies

Such inefficiencies are often fleeting and difficult to detect reliably through casual observation, since market noise and other external forces can obscure them. A systematic approach is therefore more effective. By using tools such as R to analyze large sets of historical data, traders can identify, test, and quantify these seasonal tendencies with greater consistency and rigor (Figure 3.7-1).

3.7.2 Commentary

Seasonality in financial markets refers to predictable patterns that occur at regular intervals, often due to investor behavior linked to events like corporate reporting or holidays. As these effects become widely known, traders may exploit them, diminishing their effectiveness. Thus, identifying new seasonal trends is challenging and requires rigorous statistical analysis of extensive historical data to differentiate real effects from random chance.

Listing 3.7.1: R function to analyze the end-of-month effect by computing average daily returns for each calendar day of the month.

```
library(tidyquant)
library(dplyr)
library(ggplot2)
library(lubridate)

analyze_end_of_month_effect <- function(ticker, start_date, end_date) {
  # Download stock data
  data <- tryCatch({
```

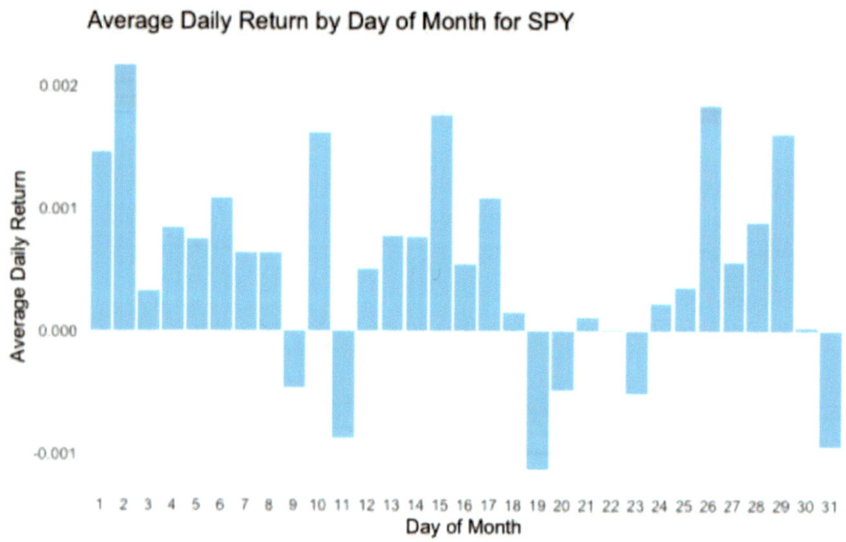

FIGURE 3.7-1: Average daily return by calendar day for SPY.

```
    tq_get(ticker, from = start_date, to = end_date, get = "stock.
    prices")
  }, error = function(e) {
    message(sprintf("Could not retrieve data for %s", ticker))
    return(NULL)
  })

  if (is.null(data) || nrow(data) == 0) {
    message(sprintf("No data available for %s", ticker))
    return(invisible(NULL))
  }
  # Compute daily returns
  returns <- data %>%
    arrange(date) %>%
    mutate(ret = adjusted / lag(adjusted) - 1) %>%
    filter(!is.na(ret)) %>%
    mutate(day_of_month = day(date))

  # Average return by day of month
  avg_returns_by_day <- returns %>%
    group_by(day_of_month) %>%
    summarise(avg_ret = mean(ret, na.rm = TRUE), .groups = "drop")

  # Plot
  ggplot(avg_returns_by_day, aes(x = factor(day_of_month), y = avg_ret)
    ) +
```

```
    geom_col(fill = "skyblue") +
    labs(
      title = paste("Average Daily Return by Day of Month for", ticker)
      ,
      x = "Day of Month",
      y = "Average Daily Return"
    ) +
    theme_minimal(base_size = 12) +
    theme(panel.grid.major.y = element_line(linetype = "dashed"))
}
# Example usage
analyze_end_of_month_effect("SPY", "2010-01-01", "2024-01-01")
```

3.8 Gaps Up/Down

Price gaps are sudden jumps in a stock's price between one day's close and the next day's open, often caused by significant news or traders adjusting valuations. They indicate temporary market inefficiencies and may present short-term opportunities. Identifying gaps is straightforward: compare the opening price to the previous day's close. A gap-filling strategy involves buying after the market opens, anticipating a price retracement to the prior close. For example, if a stock closes at \$50 and opens at \$53, that \$3 difference is a gap. If prices then decline to \$50, it's considered a gap fill. Conversely, if prices continue rising, this may signal bullish sentiment.

3.8.1 Price Gaps

Price discontinuities provide high-signal opportunities for modeling market transitions and calculating the resulting shifts in price action. Figure 3.8-1 illustrates the mechanics of overnight and intraday price gaps—specifically gap ups, gap downs, and the subsequent gap fill—which often serve as catalysts for mean-reversion or trend-extension strategies. By visualizing these imbalances between the previous close and the subsequent open, we can better understand the liquidity voids that algorithmic execution must navigate across diverse asset classes.

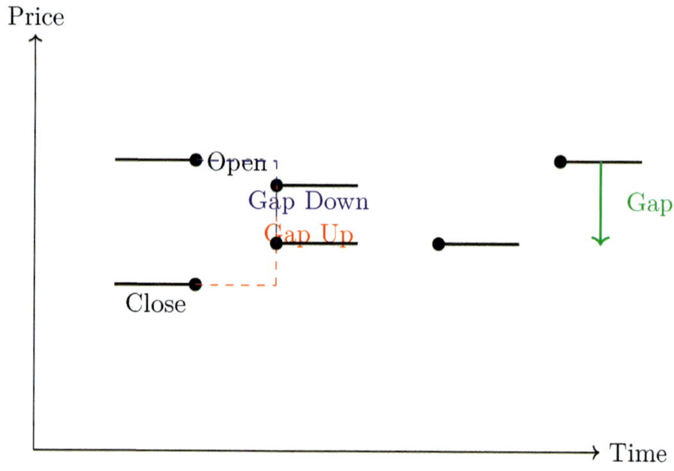

FIGURE 3.8-1: A *gap up* occurs when the opening price is significantly higher than the previous close, often on positive news. A *gap down* reflects the opposite. A *gap fill* occurs when price retraces to the original level where the gap appeared.

FIGURE 3.8-2: SPY closing prices with vertical markers indicating days where significant price gaps (exceeding 1%) occurred.

TABLE 3.2: Gap Closure Times for SPY (Threshold: 1.00%)

Gap Date	**Gap (%)**	**Direction**	**Closure Date**	**Days to Close**
2023-01-25	−1.06	Down ↓	2023-01-26	1
2023-02-03	−1.25	Down ↓	2023-05-18	104
2023-02-16	−1.25	Down ↓	2023-04-18	61
2023-02-21	−1.03	Down ↓	2023-03-31	38
2023-02-24	−1.31	Down ↓	2023-03-03	7
2023-03-13	−1.06	Down ↓	2023-03-14	1
2023-03-14	1.33	Up ↑	Open	–
2023-03-15	−1.49	Down ↓	2023-03-16	1
2023-03-29	1.10	Up ↑	Open	–
2023-11-14	1.39	Up ↑	Open	–
2024-02-13	−1.29	Down ↓	2024-02-15	2
2024-02-22	1.37	Up ↑	2024-04-19	57
2024-04-10	−1.13	Down ↓	2024-05-09	29
2024-04-25	−1.23	Down ↓	2024-04-26	1
2024-05-03	1.21	Up ↑	2025-04-07	339
2024-07-17	−1.07	Down ↓	2024-09-19	64
2024-07-31	1.29	Up ↑	2024-08-02	2
2024-08-02	−1.34	Down ↓	2024-08-14	12
2024-08-05	−3.99	Down ↓	2024-08-09	4
2024-08-07	1.21	Up ↑	2025-04-04	240
2024-08-08	1.01	Up ↑	2025-04-04	239
2024-08-15	1.06	Up ↑	2024-09-06	22
2024-09-19	1.71	Up ↑	2025-03-10	172
2024-11-06	2.17	Up ↑	2025-03-06	120
2024-12-30	−1.20	Down ↓	2025-01-06	7
2025-01-15	1.40	Up ↑	2025-03-04	48
2025-01-27	−2.17	Down ↓	2025-02-13	17
2025-02-03	−1.52	Down ↓	2025-02-05	2
2025-02-12	−1.01	Down ↓	2025-02-13	1
2025-03-06	−1.30	Down ↓	2025-05-13	68
2025-03-10	−1.45	Down ↓	2025-05-12	63
2025-03-12	1.12	Up ↑	2025-03-13	1
2025-03-21	−1.10	Down ↓	2025-03-24	3
2025-03-24	1.21	Up ↑	2025-03-28	4
2025-03-31	−1.05	Down ↓	2025-04-01	1

Continued on next page

Gap Date	Gap (%)	Direction	Closure Date	Days to Close
2025-04-02	−1.06	Down ↓	2025-05-02	30
2025-04-03	−3.44	Down ↓	2025-05-02	29
2025-04-04	−2.43	Down ↓	2025-04-09	5
2025-04-07	−3.18	Down ↓	2025-04-09	2
2025-04-08	3.47	Up ↑	Open	–
2025-04-10	−3.00	Down ↓	2025-04-25	15
2025-04-14	1.89	Up ↑	2025-04-16	2
2025-04-16	−1.10	Down ↓	2025-04-24	8
2025-04-22	1.22	Up ↑	Open	–
2025-04-23	2.50	Up ↑	Open	–
2025-04-30	−1.22	Down ↓	2025-05-01	1
2025-05-01	1.05	Up ↑	Open	–
2025-05-02	1.12	Up ↑	Open	–
2025-05-12	3.04	Up ↑	Open	–
2025-05-19	−1.03	Down ↓	2025-06-03	15
2025-05-23	−1.22	Down ↓	2025-05-27	4
2025-05-27	1.20	Up ↑	Open	–
2025-09-02	−1.17	Down ↓	2025-09-04	2

Summary of Gap Closures

Direction	Total	Closed	Open	% Open	Avg Days	Median
Down	31.00	31.00	0.00	0.00	19.29	7.00
Up	22.00	12.00	10.00	45.50	103.83	52.50
	Overall (closed only):			–	42.88	8.00

3.8.2 Commentary

Traders often say that "gaps tend to fill," meaning prices may return to the level of a gap. However, this is a statistical tendency, not a certainty. Gaps after major news can indicate a new trend (breakaway gaps) and might remain unfilled. The likelihood of filling a gap depends on its size and the market environment; larger gaps suggest a fundamental sentiment shift, while smaller gaps in stable conditions are more likely to close quickly.

Listing 3.8.1: SPY closing price gaps.

```
library(tidyquant)
library(dplyr)
library(ggplot2)
```

```
identify_price_gaps <- function(ticker, start_date, end_date, gap_threshold_percent) {
  #' Downloads stock data and identifies price gaps above a certain threshold.
  #'
  #' @param ticker Stock symbol (e.g., "SPY")
  #' @param start_date Start date (YYYY-MM-DD)
  #' @param end_date End date (YYYY-MM-DD)
  #' @param gap_threshold_percent Gap threshold in percent

  data <- tq_get(ticker, from = start_date, to = end_date)

  if (nrow(data) == 0) {
    message(sprintf("Could not retrieve data for %s.", ticker))
    return(invisible(NULL))
  }

  # Calculate gap percent: (Open - Previous Close) / Previous Close * 100
  data <- data %>%
    mutate(
      prev_close = lag(close),
      gap_percent = ((open - prev_close) / prev_close) * 100
    )

  significant_gaps <- data %>%
    filter(!is.na(gap_percent), abs(gap_percent) > gap_threshold_percent)

  # Print results
  message(sprintf("Significant Price Gaps (Threshold: %.2f%%):", gap_threshold_percent))
  apply(significant_gaps, 1, function(row) {
    gap_direction <- ifelse(as.numeric(row["gap_percent"]) > 0, "Up", "Down")
    message(sprintf("  Date: %s, Gap: %.2f%%, Direction: %s",
                    row["date"], as.numeric(row["gap_percent"]), gap_direction))
  })
  # Plot
  p <- ggplot(data, aes(x = date, y = close)) +
    geom_line(color = "blue") +
    geom_vline(data = significant_gaps, aes(xintercept = as.numeric(date)),
               color = "red", linetype = "dashed", alpha = 0.5) +
    labs(
      title = paste(ticker, "Closing Price with Significant Gaps Highlighted"),
```

```
        x = "Date",
        y = "Closing Price"
      ) +
      theme_minimal()

    print(p)
  }
  # Example usage
  identify_price_gaps("SPY", "2023-01-01", "2025-12-31", 1.0)
```

3.9 Time Series

Time-series analysis applies statistical models to data arranged in chronological order, with the goal of detecting patterns and forecasting future values. In financial markets, such models are often used to predict asset returns or the volatility of those returns. Two widely employed approaches are ARIMA (Autoregressive Integrated Moving Average) and GARCH (Generalized Autoregressive Conditional Heteroskedasticity).

3.9.1 ARIMA Models

ARIMA models attempt to forecast future values by drawing on both past values and past forecast errors in the series. GARCH models, by contrast, are designed to capture the dynamics of volatility itself, recognizing the empirical tendency for volatility to cluster—periods of heightened volatility are likely to be followed by more turbulence, while tranquil periods often persist.

A central limitation lies in the assumption that these models can consistently extract and predict non-random structure in markets that are, in practice, highly complex and often unpredictable.

3.9.2 Commentary

The plot in Figure 3.9-1 shows Microsoft's (MSFT) daily log returns from January to December 2023, along with an ARIMA-based forecast for the next trading day. The series fluctuates around zero, consistent with the random-walk nature of asset returns. Volatility clusters are visible, with several spikes early and mid-year. The ARIMA model (automatically selected using the **forecast** package) produces a one-day-ahead forecast of approximately 0.28%, with confidence intervals spanning both positive and negative values—indicating that the expected return is close to zero and largely dominated by uncertainty. This highlights the difficulty of predicting short-term returns using ARIMA models alone.

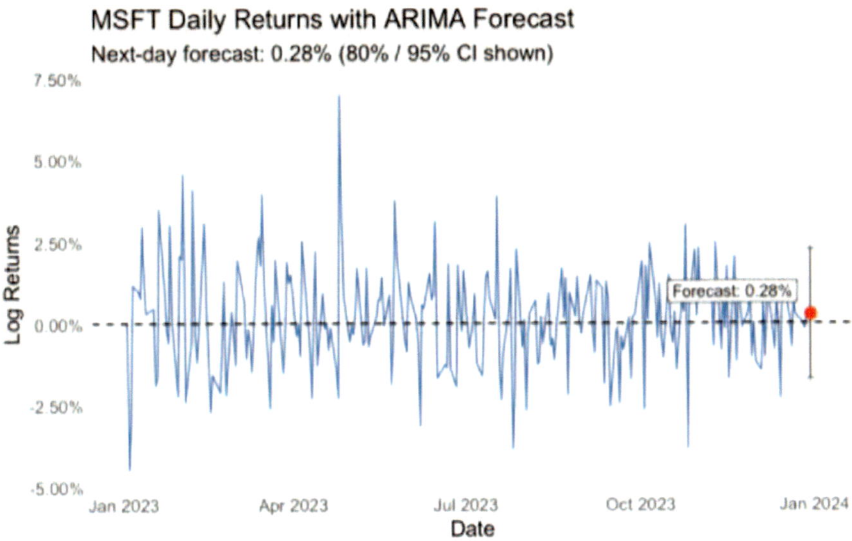

FIGURE 3.9-1: MSFT daily log returns in 2023 with ARIMA forecast.

Listing 3.9.1: MSFT daily log returns in 2023 with ARIMA forecast.

```
# ---- Average Daily Return by Day-of-Month (R version)
    -----------------------
# Requirements: tidyquant, dplyr, ggplot2, lubridate, scales
suppressPackageStartupMessages({
  library(tidyquant)
  library(dplyr)
  library(ggplot2)
  library(lubridate)
  library(scales)
})

analyze_end_of_month_effect <- function(ticker, start_date, end_date) {
  # Downloads daily Close prices and computes average returns by day of
     month.
  # Args:
  #   ticker: e.g., "SPY"
  #   start_date, end_date: "YYYY-MM-DD"
  # Returns:
  #   A tibble with columns: day, avg_return (in decimal). Also prints
    a bar chart.

  prices <- tq_get(ticker, get = "stock.prices",
                   from = start_date, to = end_date, complete_cases =
    TRUE)
```

```
  if (nrow(prices) == 0) {
    stop(sprintf("Could not retrieve data for %s.", ticker))
  }

  df <- prices %>%
    arrange(date) %>%
    transmute(date, close = close) %>%
    mutate(ret = close / dplyr::lag(close) - 1) %>%
    filter(!is.na(ret)) %>%
    mutate(day = lubridate::day(date))

  avg_by_day <- df %>%
    group_by(day) %>%
    summarise(avg_return = mean(ret, na.rm = TRUE), .groups = "drop")

  p <- ggplot(avg_by_day, aes(x = factor(day), y = avg_return)) +
    geom_col() +
    labs(
      title = paste0("Average Daily Return by Day of Month for ",
    ticker),
      subtitle = paste0(format(as.Date(start_date), "%Y-%m-%d"),
                        " to ", format(as.Date(end_date), "%Y-%m-%d")),
      x = "Day of Month",
      y = "Average Daily Return"
    ) +
    scale_y_continuous(labels = scales::percent_format(accuracy = 0.01)
    ) +
    theme_minimal(base_size = 13) +
    theme(panel.grid.major.x = element_blank())

  print(p)
  invisible(avg_by_day)
}

# ---- Example usage
    -------------------------------------------------------------
avg_tbl <- analyze_end_of_month_effect("SPY", "2015-01-01", "2025-09-07
    ")
head(avg_tbl)
```

3.10 Price Shocks

Significant news events—such as company earnings reports or the release of unexpected economic data—can produce sudden and substantial price movements known as price shocks. These sharp moves may create a temporary market inefficiency if traders can detect the early signs of strong momentum and act quickly to exploit it.

3.10.1 Relative Strength Index

One tool for monitoring such momentum is the Relative Strength Index (RSI). The RSI measures the speed and magnitude of recent price changes on a standardized scale. A sharp rise in price, particularly when accompanied by a rapidly climbing RSI, often signals strong upward momentum. In the context of a price shock, this pattern suggests that the move may persist for a short period, creating opportunities for well-timed trades.

FIGURE 3.10-1: GME closing price (blue line, left axis) with Relative Strength Index (RSI, orange dashed line, right axis).

Figure 3.10-1 displays potential price shocks highlighted as red circles, defined as days when the price change exceeded 5% while the RSI was above 70. The clustering of shocks in May 2024 corresponds to an extreme speculative surge, while later shocks in 2025 reflect smaller but still notable overbought conditions.

3.10.2 Commentary

Table 3.3 provides the precise dates, magnitudes of price changes, and RSI values corresponding to the potential shocks highlighted as red circles in Figure 3.10-1. Together, they illustrate how extreme short-term moves align with periods of elevated momentum.

TABLE 3.3: Potential price shocks (price change > 5%, RSI > 70).

Date	Price Change (%)	RSI
2023-03-22	35.24	70.95
2023-06-07	5.75	70.65
2024-05-03	29.08	74.14
2024-05-09	13.13	75.89
2024-05-13	74.40	88.24
2024-05-14	60.10	93.77
2024-05-15	−18.87	74.75
2024-06-06	47.45	72.11
2024-11-11	9.57	77.77
2024-11-25	6.34	74.22
2025-05-23	7.03	75.10
2025-05-27	5.99	79.12

1. Spring 2024 cluster (May 13–15, 2024): The table confirms three consecutive shocks: +74.40%, +60.10%, and −18.87% in just three sessions. The RSI soared to extreme levels (93.77 at the peak), corroborating the speculative surge seen in the figure. This period represents a textbook case of overextension followed by rapid correction.

2. Moderate shocks (2023 and late 2024–2025): Smaller shocks (e.g., +5.75% on June 7, 2023; +9.57% on November 11, 2024) occur when RSI hovers in the 70–80 range. These align with local tops in the figure but did not trigger as dramatic a reversal as the May 2024 cluster. Instead, they represent overbought conditions where traders should exercise caution but not necessarily expect immediate collapse.

3. Exceptionally large positive shocks (March 22, 2023 and June 6, 2024): The table shows jumps of +35.24% and +47.45%, both with RSI only moderately over 70. In Figure 3.10-1, these shocks coincide with sudden spikes, often the result of news-driven momentum.

Strategic implications:

The combined figure and table results show that the magnitude of price shocks increases with clustering—isolated shocks which may fade gradually, but clusters (like May 2024) mark high-risk turning points. Monitoring RSI alone is insufficient; it is the co-occurrence of extreme RSI and consecutive outsized price changes that signals instability.

Listing 3.10.1: Script for detecting and visualizing potential price shocks in GME

```
# Required packages
library(quantmod)
library(TTR)
library(dplyr)
library(ggplot2)

detect_price_shocks <- function(ticker, start_date, end_date, price_
    change_threshold_percent, rsi_threshold) {
  # Download stock data
  getSymbols(ticker, src = "yahoo", from = start_date, to = end_date,
    auto.assign = TRUE, warnings = FALSE)
  data <- get(ticker)

  if (nrow(data) == 0) {
    cat(sprintf("Could not retrieve data for %s.\n", ticker))
    return(NULL)
  }

  # Convert to data frame
  df <- data.frame(date = index(data), coredata(data))
  colnames(df) <- c("Date", "Open", "High", "Low", "Close", "Volume", "
    Adjusted")

  # Compute price change % and RSI
  df <- df %>%
    mutate(PriceChange = (Close / lag(Close) - 1) * 100,
           RSI = RSI(Close, n = 14))

  # Detect shocks
  df <- df %>%
    mutate(Shock = (abs(PriceChange) > price_change_threshold_percent)
    & (RSI > rsi_threshold))

  shocks <- df %>% filter(Shock)

  cat(sprintf("Potential Price Shocks (Price Change > %.1f%%, RSI > %.1
    f):\n",
              price_change_threshold_percent, rsi_threshold))
  for (i in seq_len(nrow(shocks))) {
    cat(sprintf("  Date: %s, Price Change: %.2f%%, RSI: %.2f\n",
                format(shocks$Date[i], "%Y-%m-%d"),
                shocks$PriceChange[i],
                shocks$RSI[i]))
  }

  # Scale RSI to overlay on price axis
  price_range <- range(df$Close, na.rm = TRUE)
```

```
  scale_factor <- diff(price_range) / 100  #since RSI ranges 0-100

  # Plot price with shocks and RSI as secondary axis
  ggplot(df, aes(x = Date)) +
    geom_line(aes(y = Close), color = "blue") +
    geom_point(data = shocks, aes(y = Close), color = "red", size = 2)
    +
    geom_line(aes(y = RSI * scale_factor + min(price_range)), color = "
    orange", linetype = "dashed") +
    scale_y_continuous(
      name = "Closing Price",
      sec.axis = sec_axis(~ (. - min(price_range)) / scale_factor,
                          name = "RSI (0-100)")
    ) +
    labs(title = paste0(ticker, " Closing Price with Potential Price
    Shocks Highlighted"),
         x = "Date") +
    theme_minimal()
}

# Example usage
detect_price_shocks("GME", "2023-01-01", "2025-09-30", 5.0, 70)
```

3.11 Volatility Breakout

The Volatility Breakout strategy seeks to take advantage of sharp price movements that occur when market volatility rises above a critical threshold. The central idea is that heightened volatility often signals the potential for significant price swings. Traders can attempt to profit by entering positions in the direction of these breakouts.

3.11.1 Volatility Breakout Signals

The R script accompanying this strategy illustrates how such signals can be identified. It begins by importing stock price data and calculating a 20-day rolling standard deviation as a measure of volatility. A predefined threshold is then applied: when volatility surpasses this benchmark, the script generates a buy signal, anticipating a strong upward move. Conversely, when volatility falls below the threshold, the script issues a sell signal, reflecting expectations of weaker price movement. In this way, the script demonstrates how systematic rules based on volatility can be used to capture potential trading opportunities.

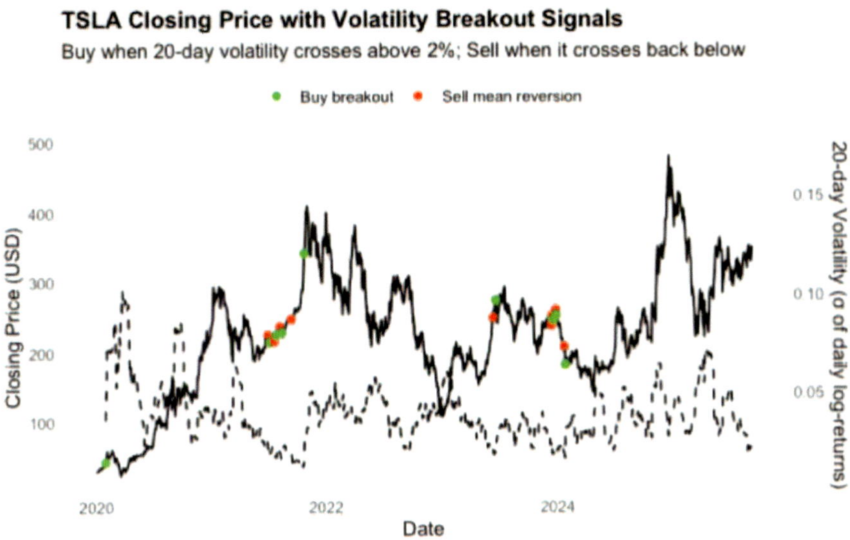

FIGURE 3.11-1: TSLA closing price (solid black line, left axis) with 20-day rolling volatility of daily log-returns (dashed line, right axis).

3.11.2 Commentary

Figure 3.11-1 displays a volatility breakout buy signal (green marker) generated when volatility crosses above the 2% threshold, signaling entry into a high-volatility regime often associated with strong price moves. Conversely, a a volatility breakout sell signal (red marker) is triggered when volatility falls back below the 2% threshold, indicating a return to a low-volatility regime and suggesting an exit. This regime-based approach highlights how systematic rules tied to volatility shifts can guide trading decisions.

Listing 3.11.1: Volatility breakout.

```
library(quantmod)
library(dplyr)
library(zoo)

# Load TSLA stock data from Yahoo Finance
getSymbols("TSLA", src = "yahoo", from = "2020-01-01", auto.assign =
    TRUE)
data <- TSLA

# Use the adjusted close price
data <- data %>%
  as.data.frame() %>%
  mutate(Date = index(TSLA),
         Close = TSLA.Adjusted)
```

```
# Calculate historical volatility (20-day rolling standard deviation)
data <- data %>%
  mutate(
    Historical_Volatility = rollapply(Close, width = 20, FUN = sd, fill
      = NA, align = "right")
  )

# Define a volatility threshold (e.g., 0.02 or 2%)
volatility_threshold <- 0.02

# Generate buy and sell signals based on volatility breakouts
data <- data %>%
  mutate(
    Signal = case_when(
      Historical_Volatility > volatility_threshold ~ 1,   # Buy signal
      Historical_Volatility < volatility_threshold ~ -1,  # Sell signal
      TRUE ~ 0                                            # Neutral
    )
  )

head(data, 25)  # View first rows with volatility and signals

# TSLA Volatility Breakout (R)
# install.packages(c("quantmod", "dplyr", "zoo", "ggplot2"))
library(quantmod)
library(dplyr)
library(zoo)
library(ggplot2)

# 1) Download TSLA from Yahoo
getSymbols("TSLA", src = "yahoo", from = "2020-01-01", auto.assign =
    TRUE)
px <- Ad(TSLA)                                  # Adjusted close (xts)
df <- data.frame(
  Date  = index(px),
  Close = as.numeric(px)
)

# 2) Daily returns and 20-day rolling volatility (sd of returns)
df <- df %>%
  mutate(
    ret = c(NA, diff(log(Close))),              # log returns
    vol20 = rollapply(ret, 20, sd, fill = NA, align = "right")
  )

# 3) Threshold & signals
volatility_threshold <- 0.02                     # approx 2% daily-
    returns volatility
```

```
df <- df %>%
  mutate(
    Signal = case_when(
      is.na(vol20) ~ 0L,
      vol20 > volatility_threshold ~ 1L,       # breakout / high-vol
    regime
      vol20 < volatility_threshold ~ -1L,      # low-vol regime
      TRUE ~ 0L
    ),
    SignalLag = dplyr::lag(Signal, default = 0L),
    # Mark only *crossovers* to reduce clutter:
    Event = case_when(
      SignalLag <= 0 & Signal == 1  ~ "Buy breakout",           # crossed
     up
      SignalLag >= 0 & Signal == -1 ~ "Sell mean reversion",  # crossed
     down
      TRUE ~ NA_character_
    )
  )

# 4) Scale volatility for secondary axis
#    Map vol20 to the price scale, then show the original units on sec.
    axis.
scale_factor <- max(df$Close, na.rm = TRUE) / max(df$vol20, na.rm =
    TRUE) * 0.6
df <- df %>% mutate(vol20_scaled = vol20 * scale_factor)

# 5) Plot
p <- ggplot(df, aes(x = Date)) +
  # Price
  geom_line(aes(y = Close), linewidth = 0.7, alpha = 0.9, na.rm = TRUE)
     +
  # Volatility on secondary axis (scaled)
  geom_line(aes(y = vol20_scaled),
            linewidth = 0.7, linetype = "dashed", alpha = 0.9, na.rm =
    TRUE) +
  # Event markers on the price series
  geom_point(
    data = subset(df, !is.na(Event)),
    aes(y = Close, color = Event, shape = Event),
    size = 2.5, stroke = 0.2, na.rm = TRUE
  ) +
  scale_color_manual(values = c("Buy breakout" = "green",
                                "Sell mean reversion" = "red")) +
  scale_shape_manual(values = c("Buy breakout" = 16, "Sell mean
    reversion" = 16)) +
  scale_y_continuous(
    name = "Closing Price (USD)",
    sec.axis = sec_axis(~ . / scale_factor,
```

```
                            name = "20-day Volatility ( sigma of daily log-returns)")
  ) +
  labs(
    title = "TSLA Closing Price with Volatility Breakout Signals",
    subtitle = "Buy when 20-day volatility crosses above 2%; Sell when it crosses back below",
    x = "Date", y = "Closing Price"
  ) +
  theme_minimal(base_size = 12) +
  theme(
    plot.title    = element_text(face = "bold", size = 14),
    legend.title  = element_blank(),
    legend.position = "top"
  )

print(p)
```

3.12 Machine Learning-Based

Machine learning–based trading strategies rely on predictive models to guide buy, sell, or hold decisions. In this example, we introduce a simple decision tree classifier as the predictive model. Decision trees are especially useful because they are both versatile and easy to interpret. They allow traders to classify outcomes—such as whether to take a long position, exit, or remain in the market—based on observed patterns in the data.

3.12.1 Decision Tree Classifier

The following R script (Listing lst:mlb) illustrates how a machine learning model can be applied to develop a trading strategy. The script begins by loading historical stock data, from which it extracts relevant features, such as technical indicators. These features are paired with corresponding trading signals that serve as labels. The dataset is then divided into training and testing subsets. A decision tree classifier is trained on the training set to learn the relationships between features and trading signals. Finally, the trained model generates predictions on the testing set, enabling us to evaluate its performance under conditions that approximate real-world trading scenarios.

3.12.2 Commentary

In Figure 3.12-1, buy signals (green markers) and sell signals (red markers) are generated when volatility shifts regimes and at least 30 trading days have

FIGURE 3.12-1: SPY closing prices (black line, left axis) with 20-day annualized volatility (orange dashed line, right axis).

elapsed since the last trade, with the trigger condition defined as an absolute next-day log return of at least 1.0%. The dashed horizontal line indicates the 20% volatility threshold used to distinguish regimes.

Table 3.4 provides a chronological audit of trading signals generated under the decision tree classifier, specifically tracking the transition between long and short positions alongside key volatility metrics. Each entry records the asset's closing price at the time of the signal, paired with the 30-day annualized volatility to provide context on the market regime during execution. This longitudinal view is essential for evaluating the stability of the signal generation logic across diverse market cycles—from the high-volatility periods of late 2015 to the more recent projections in 2025. By documenting these transitions, the table facilitates a precise calculation of volatility associated with specific price movements following each Buy or Sell trigger, ensuring that risk management remains calibrated to shifting variance levels.

TABLE 3.4: Trading signals with closing price and 30-day annualized volatility.

Date	**Signal**	**Close**	**Vol. Ann. (%)**
2015-08-26	Buy	163.92	21.84
2015-10-14	Sell	168.87	18.30
2016-01-22	Buy	162.40	20.54
2016-03-07	Sell	170.98	19.36
2018-02-08	Buy	228.68	20.62
2018-04-23	Sell	237.56	19.75
2018-11-06	Buy	247.40	20.12

Continued on next page...

Date	Signal	Close	Vol. Ann. (%)
2019-08-28	Buy	263.73	20.03
2020-09-24	Sell	302.42	19.85
2022-04-25	Sell	408.93	19.91
2022-09-08	Sell	383.73	19.14
2025-04-03	Buy	535.12	23.76
2025-05-23	Sell	577.40	19.88

Listing 3.12.1 implements a machine learning-based approach for generating stock trading signals using a Decision Tree Classifier.

Listing 3.12.1: Decision tree classifier.

```
# ---- Packages ----
suppressPackageStartupMessages({
  library(quantmod)  # data + indicators (via TTR)
  library(rpart)     # decision tree
})

# ---- 1) Download SPY data from Yahoo ----
getSymbols("SPY", src = "yahoo", from = "2015-01-01", auto.assign =
    TRUE)
df <- na.omit(data.frame(
  Date   = index(SPY),
  Open   = as.numeric(Op(SPY)),
  High   = as.numeric(Hi(SPY)),
  Low    = as.numeric(Lo(SPY)),
  Close  = as.numeric(Cl(SPY)),
  Volume= as.numeric(Vo(SPY))
))

# ---- 2) Create features (example technical indicators) ----
# Feature1: 14-day RSI
df$RSI14 <- RSI(df$Close, n = 14)

# Feature2: Distance of price from 20-day SMA (as %)
sma20 <- SMA(df$Close, n = 20)
df$SMA20_DistPct <- 100 * (df$Close - sma20) / sma20

# Feature3: MACD histogram (12,26,9) - momentum proxy
macd <- MACD(df$Close, nFast = 12, nSlow = 26, nSig = 9)
df$MACD_Hist <- macd[, "macd"] - macd[, "signal"]

# Keep only rows with all features present
df <- na.omit(df)

# ---- 3) Create labels (trading signals) ----
# Label rule:
#   +1 (Buy)  if next-day return > +0.5%
```

```
#    0 (Hold) if in [-0.5%, +0.5%]
#   -1 (Sell) if next-day return < -0.5%
ret_next <- c(diff(log(df$Close)), NA)          # next-day log return
thr <- 0.005                                    # 0.5% threshold in log-
    return terms (~0.5%)
signal <- ifelse(ret_next >  thr,  1L,
                 ifelse(ret_next < -thr, -1L, 0L))
df$Signal <- signal

# Drop last row (no label for final day)
df <- df[!is.na(df$Signal), ]

# ---- 4) Define features (X) and label (y) ----
X <- df[, c("RSI14", "SMA20_DistPct", "MACD_Hist")]
y <- factor(df$Signal, levels = c(-1, 0, 1), labels = c("Sell", "Hold",
     "Buy"))
data_model <- data.frame(X, Signal = y)

# ---- 5) Train/Test split (80/20, time-order aware to avoid leakage)
    ----
set.seed(42)
n <- nrow(data_model)
split_idx <- floor(0.8 * n)
train <- data_model[1:split_idx, ]
test  <- data_model[(split_idx + 1):n, ]

# ---- 6) Train decision tree ----
tree_fit <- rpart(Signal ~ ., data = train, method = "class", control =
     rpart.control(cp = 0.001))

# ---- 7) Predict on test set ----
pred <- predict(tree_fit, newdata = test, type = "class")

# ---- 8) Simple evaluation ----
cm <- table(Actual = test$Signal, Predicted = pred)
acc <- mean(pred == test$Signal)

print(cm)
cat(sprintf("\nAccuracy: %.2f%%\n", 100 * acc))
```

Listing 3.12.1 begins by loading historical price data and associated features using **quantmod**. The features, representing technical indicators, are stored in X, while the trading signals, which indicate whether to buy (1), sell (-1), or hold (0) a stock, are stored in y. The data is then divided into training and testing sets, allocating 80% for training and 20% for testing, with a specific random seed for consistency. A Decision Tree Classifier is created and trained using the training data. Subsequently, the model is used to make predictions on the testing set, generating trading signals based on the learned patterns in the historical data. This approach leverages machine learning to automate the decision-making process for stock trading, potentially aiding investors in making informed decisions based on historical features and patterns.

4

Performance Testing

Dare to live the life you have dreamed for yourself. Go forward and make your dreams come true.

(Ralph Waldo Emerson)

4.1 Backtesting with Historical Data I

Backtesting is a systematic method for evaluating a trading strategy using historical data before risking real money. Traders apply defined rules to past price movements to assess how a strategy would have performed over a specific period, based on the idea that historical patterns can provide insights into future market behavior. This process helps identify the strengths and weaknesses of a strategy, refine parameters, and build confidence in its potential. In algorithmic trading, backtesting is an essential step before deploying any model.

The process begins with a clearly defined trading strategy and specific rules. Historical data that closely simulates real trading conditions is gathered, and the strategy is executed on this dataset according to the established rules. Results are then evaluated using performance metrics like the Sharpe ratio and maximum drawdown to assess effectiveness and risk.

The reliability of backtesting depends on data quality, making careful preparation crucial. Historical datasets must be accurate, consistent, and free from biases. This involves cleaning the data, addressing missing observations, and normalizing values as needed.

Accurate backtesting requires reliable historical data. High-quality data can be obtained from reputable providers such as Interactive Brokers (IBKR), the Bloomberg Terminal, or more accessible options like Quandl and Alpha Vantage.

Common challenges include managing missing values, which can distort results if not addressed. Methods such as linear interpolation and forward or backward filling can replace missing points, while more advanced techniques like multiple imputation can estimate values based on relationships within the data.

Normalization and transformation are also important to ensure the dataset is suitable for backtesting. Techniques like Min–Max scaling and logarithmic adjustments can stabilize variance and improve alignment with the assumptions of a trading strategy.

Lastly, maintaining data integrity and quality is critical. Traders should check for outliers, identify anomalies, and ensure that datasets match the intended scope. Cross-referencing with trusted sources bolsters confidence in data reliability for backtesting.

4.1.1 Introduction

Why Use R for Backtesting?

Backtesting is a vital component in the development of a successful trading strategy. It empowers traders to identify potential weaknesses and make decisive, informed choices about the effectiveness of their concepts. By employing backtesting, traders can refine their strategies and leverage data-driven insights, significantly increasing their chances of success.

Why Not Use R for Backtesting?

R is a powerful tool, but it isn't suitable for everyone due to its steep learning curve, especially for those unfamiliar with programming or statistics. Many traders may find that alternative software better meets their specific data analysis and visualization needs.

Prerequisites

This section delivers straightforward and effective code, and it's essential that you possess a basic understanding of programming and R, along with fundamental trading strategies. Our primary focus will be on R and backtesting.

The Comprehensive R Archive Network (CRAN) is a vital resource for R, offering a vast array of useful packages. You must install two specific packages for this script: **quantmod** and **PerformanceAnalytics**.

Listing 4.1.1: Prerequisite packages for back testing in R.

```
suppressPackageStartupMessages({
  library(quantmod) # data + indicators (via TTR)
  library(TTR)
  library(PerformanceAnalytics) # decision tree
})

library(quantmod)
library(TTR)
library(PerformanceAnalytics)
```

4.1.2 Backtesting in R

This section directly builds upon Eric Goebelbecker's excellent tutorial, effectively correcting bugs and significantly enhancing the performance analytics.

You can break the backtesting process down into four essential steps:

1. Retrieve historical price data.
2. Use the data to create technical indicators.
3. Apply the indicators to the historical prices.
4. Back test the strategy to see how it performs.

Download Historical Price Data

Listing 4.1.2: Retrieve historical price data.

```
get_and_graph_prices <- function(ticker, from = "2024-01-01", to = Sys.
    Date()) {

  # Download data from Yahoo Finance
  prices <- getSymbols(ticker, from = from, to = to, auto.assign =
    FALSE)

  # Graph them in a bar chart
  barChart(prices, name = ticker, bar.type = "hlc", up.col = "green",
    dn.col = "red", theme = "white")

  return(prices)
}
```

The function in Listing 4.1.2 accepts a ticker symbol and optional starting and ending dates for the data. **getSymbols()** returns a data frame with columns based on the symbol type. So, calling **get_and_graph_prices** with TSLA gives Figure 4.1-1.

Listing 4.1.3: Retrieve TSLA historical price data.

```
prices <- get_and_graph_prices("TSLA")
head(prices,n=5)
```

Figure 4.1-2 shows the first five rows of an `xts` object in R. Here, the dates (e.g., 2024-01-02, 2024-01-03, etc.) serve as the index, while the associated OHLCV (Open, High, Low, Close, Volume, Adjusted) values of TSLA are stored as the data aligned to that index. This index–data structure is what makes `xts` especially suited for time series financial data.

However, we may sometimes want to extract the index as a `Date` object. Doing so allows us to:

- Use the dates in plots or as labels.

FIGURE 4.1-1: Candle chart of TSLA daily prices and trading volume.

- Perform date arithmetic, such as calculating time differences between observations.
- Merge the financial data with other datasets that use explicit date columns (e.g., macroeconomic indicators).
- Filter or group the data by components of the date (e.g., month, quarter, year).

This gives us greater flexibility in downstream analysis and makes the dates accessible as a regular variable rather than hidden in the xts index.

The following R code extracts the index as a `Date` object.

```
head(prices,n=5)
```

```
##            TSLA.Open TSLA.High TSLA.Low TSLA.Close TSLA.Volume TSLA.Adjusted
## 2024-01-02    250.08    251.25   244.41     248.42   104654200        248.42
## 2024-01-03    244.98    245.68   236.32     238.45   121082600        238.45
## 2024-01-04    239.25    242.70   237.73     237.93   102629300        237.93
## 2024-01-05    236.86    240.12   234.90     237.49    92488900        237.49
## 2024-01-08    236.14    241.25   235.30     240.45    85166600        240.45
```

FIGURE 4.1-2: TSLA daily closing prices and trading volume.

Listing 4.1.4: Extract the index of an `xts` object as `Date` values for use in analysis and plotting.

```
# 1. Extract index as Date
dates <- index(prices)
head(dates)

# 2. Convert to a data.frame with Date column
prices_df <- data.frame(Date = index(prices), coredata(prices))
head(prices_df)

# 3. If you only need rownames as string dates
date_strings <- rownames(prices)
head(date_strings)
```

We can use the prices data frame, `prices_df`, to calculate the technical indicators.

Creating Indicators

We will calculate a simple moving average (SMA) for our technical indicator. This indicator is straightforward and may not provide significant results, but our primary focus is on backtesting rather than developing a trading strategy.

quantmod gives us `SMA()` for calculating SMAs. The **TTR** package includes an impressive set of functions that you can mix and match to form your own strategy.

`SMA()` requires a data frame of prices and a specified interval. However, `getSymbols()` returns a named list that unfortunately incorporates the ticker names within the column names. For instance, if we were running a one-time script for TSLA, we would proceed as follows:

Listing 4.1.5: TSLA with 20-day simple moving average (SMA).

```
# Close any broken devices and start clean
try(graphics.off(), silent = TRUE)

suppressPackageStartupMessages({ library(quantmod); library(TTR) })
options("getSymbols.warning4.0" = FALSE)

getSymbols("TSLA", from = "2023-01-01", src = "yahoo", auto.assign =
    TRUE)
sma20 <- SMA(Cl(TSLA), n = 20)
chartSeries(TSLA, name = "TSLA -- SMA(20)", TA = NULL)  # no theme
addTA(sma20, on = 1, col = "gold", lwd = 2, legend = "SMA(20)")
```

Listing 4.1.5 gives Figure 4.1-3. Figure 4.1-4 presents the **quantmod** version with BUY and SELL triangle markers overlaid on the price chart using `addTA()`:

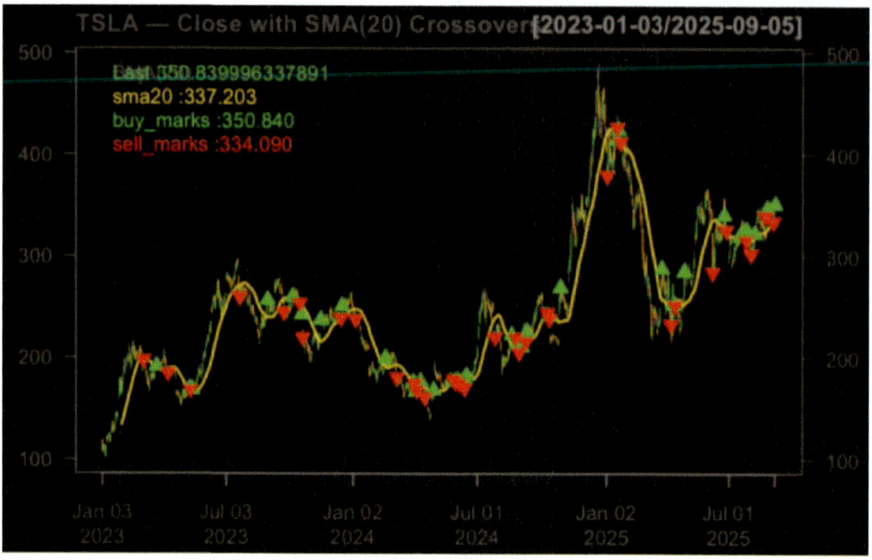

FIGURE 4.1-3: TSLA with 20-day SMA (gold line).

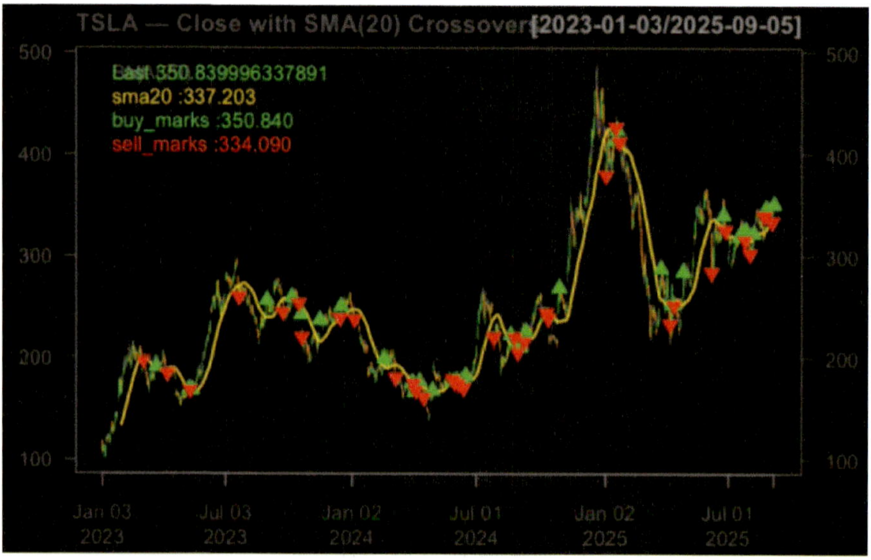

FIGURE 4.1-4: TSLA with 20-day SMA (gold line) with **BUY** and **SELL** triangle markers.

Listing 4.1.6: TSLA script with 20-day SMA (gold line) with **BUY** and **SELL** triangle markers.

```
suppressPackageStartupMessages({ library(quantmod); library(TTR)})
options("getSymbols.warning4.0" = FALSE)

# --- Data
getSymbols("TSLA", from = "2023-01-01", src = "yahoo", auto.assign =
    TRUE)
price <- Cl(TSLA)
sma20 <- SMA(price, n = 20)

# --- Crossover detection (price vs SMA)
state <- ifelse(price > sma20, 1, -1)          # 1=above, -1=below
cross <- diff(state)                           # +2 = cross up (buy), -2
     = cross down (sell)

buy_idx  <- which(cross ==  2) + 1
sell_idx <- which(cross == -2) + 1

# Marker series (xts with NAs except at crossover points)
buy_marks  <- xts(rep(NA_real_, NROW(price)), index(price))
sell_marks <- xts(rep(NA_real_, NROW(price)), index(price))
buy_marks[buy_idx]    <- price[buy_idx]
sell_marks[sell_idx] <- price[sell_idx]

# --- Chart (no custom theme to avoid device/theme bugs)
chartSeries(TSLA, name = "TSLA - Close with SMA(20) Crossovers", TA =
    NULL, legend.loc = "topleft")
addTA(sma20, on = 1, col = "gold", lwd = 2, legend = "SMA(20)")

# Buy: triangle up (pch=24), green fill
addTA(buy_marks,  on = 1, type = "p", pch = 24, col = "green", bg = "
    green", cex = 1.2, legend = "Buy")

# Sell: triangle down (pch=25), red fill
addTA(sell_marks, on = 1, type = "p", pch = 25, col = "red", bg = "red"
    , cex = 1.2, legend = "Sell")
```

The chart shown in Figure 4.1-4 presents Tesla's daily closing price (TSLA) from January 2023 through September 2025, overlaid with a 20-day Simple Moving Average (SMA 20) and annotated buy and sell crossover signals.

Key Elements:

Yellow Line (SMA 20): smoothing of the last 20 trading days' closing prices.

Green Up Triangles (Buy Marks): Indicate points where the price crossed above the SMA 20, signaling short-term bullish momentum.

Red Down Triangles (Sell Marks): Indicate points where the price crossed below the SMA 20, signaling potential bearish momentum.

Legend Values: Indicate points where the price crossed below the SMA 20, signaling potential bearish momentum.

Last: Latest TSLA price (≈350.84).

sma20: Most recent SMA 20 value (≈337.20).

buy_marks: Last buy signal price (≈350.84).

sell_marks: Last sell signal price (≈334.09).

Interpretation of SMA 20 Crossovers:

1. **Trend Following Logic**: When TSLA closes above SMA 20, the system generates a buy signal (green marker), anticipating continuation of upward momentum. When TSLA closes below SMA 20, a sell signal is triggered (red marker), anticipating potential decline.
2. **Observed Period Behavior**: 2023–mid 2024: Frequent crossovers, creating many signals. This reflects choppy trading and sideways movement where SMA crossovers tend to produce false or short-lived signals.
3. **Late 2024–early 2025**: A strong rally with sustained price action above SMA 20. Buy signals dominated, showing the SMA20 effectively caught the uptrend.
4. **Early 2025 crash**: Sharp breakdown where sell signals clustered as TSLA fell back under SMA 20.
5. **Mid 2025**: Volatility and oscillation with alternating buy and sell signals, showing indecision and consolidation.
6. **Signal Reliability**: The SMA 20 is a short-term momentum filter which is responsive but noisy. It works best in trending markets (e.g., late 2024 rally), but produces whipsaws in sideways conditions (e.g., early 2023 and mid 2025).

Strategic Takeaway:

- The 20-day SMA crossover is useful for capturing medium-term trends in TSLA, but not reliable as a stand-alone system in ranging conditions.
- Confirm with a longer SMA (e.g., 50- or 100-day) for trend direction.
- Add volume or volatility filters to reduce false signals.
- Require a holding period (e.g., 30 trading days) to avoid overtrading.

In summary

This chart (Figure 4.1-4) illustrates how the SMA 20 crossover captures TSLA's big trend shifts (late 2024 uptrend, early 2025 downtrend) but also shows the method's weakness—excess signals in sideways markets.

4.1.3 Performance Backtest

In this subsection, we backtest the SMA 20 crossover strategy on TSLA (2023–2025) to illustrate its performance in terms of returns, Sharpe ratio, and drawdowns.

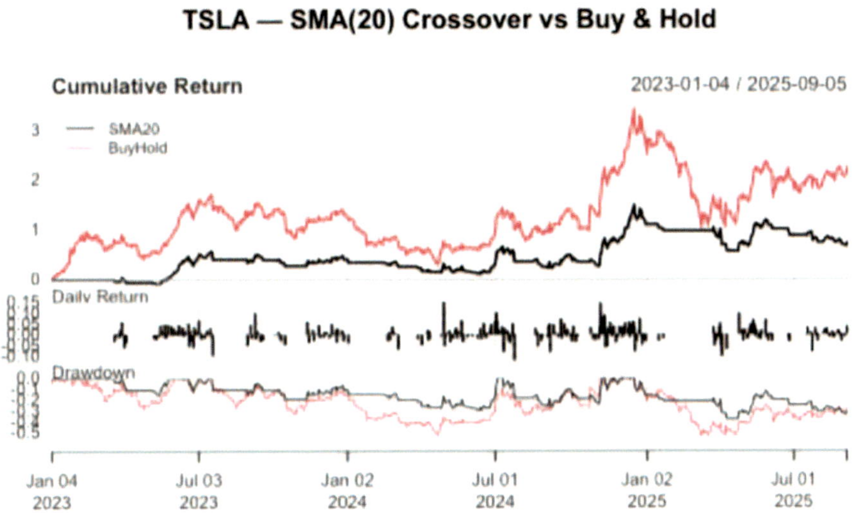

FIGURE 4.1-5: Performance analysis of TSLA 20SMA crossover.

The graph in Figure 4.1-5 compares TSLA's performance using a 20-day simple moving average (SMA 20) crossover strategy versus a simple buy and hold benchmark over the period 2023-01-04 to 2025-09-05. It contains three panels: cumulative return, daily return, and drawdown (from top to bottom).

Cumulative Return

Buy and hold (red line) clearly outperformed the SMA 20 strategy. It grew to more than 3× the initial investment at its peak (early 2025), before pulling back but still ending near 2.5×. This reflects Tesla's strong bull run in 2024–2025, which rewarded staying continuously invested. The SMA 20 strategy (black line) remained flat to modestly positive, ending just slightly above break-even, 1.2×. The SMA 20 crossover system frequently exited and re-entered the market, which limited gains during prolonged uptrends. By avoiding full exposure, it also missed the steepest part of the bull rally.

Daily Return

The SMA 20 returns are more sporadic, with exposure only during crossover-signaled periods. Returns appear smaller in magnitude compared to buy and hold. Buy and hold generated larger swings in daily returns, particularly visible during 2024–2025 when TSLA's volatility surged. Interpretation: the SMA 20 lowered volatility and exposure, but at the expense of failing to capture major upside.

Drawdown

Buy and hold (red line) experienced sharper drawdowns, particularly in mid-2025 after the rally peaked. Losses from the top exceeded 30% before partial recovery. The SMA 20 (black line) had shallower drawdowns, generally capped below 10–15%. This reflects the defensive nature of the strategy: exiting when price crosses below the SMA 20 helps limit downside.

Strategic Trade-offs

Strengths of the SMA 20 Crossover Strategy:

- Lower volatility, lower drawdowns.
- Preserves capital in downturns.
- More risk-averse: better suited for conservative traders prioritizing downside protection.
- Require a holding period (e.g., 30 trading days) to avoid overtrading.

Weaknesses of the SMA 20 Crossover Strategy:

- Whipsaws in sideways markets led to flat returns.
- Missed extended uptrends (e.g., 2024–2025 bull run).

Strengths of Buy and Hold:

- Maximum exposure to strong secular rallies.
- Simpler execution.

Weaknesses of Buy and Hold:

- Higher volatility, deeper drawdowns.
- Requires investor tolerance for large swings.

Overall Takeaway

For this period, Buy and Hold massively outperformed the SMA(20) crossover strategy in absolute returns. The SMA(20) provided smoother equity curves with significantly smaller drawdowns, but at the cost of underperformance. SMA(20) acted as a risk-reduction filter, but Tesla's trend structure in 2023–2025 favored uninterrupted exposure rather than timing exits.

Summary

- If the goal was maximum return, Buy and Hold dominated.
- If the goal was capital preservation and lower drawdowns, the SMA(20) offered stability but sacrificed upside.

The backtest results are presented in Table 4.1.

TABLE 4.1: Backtest results: entry and exit trades with holding periods and returns.

Entry Date	Entry Price	Exit Date	Exit Price	Hold Days	Return
2023-03-21	197.58	2023-04-05	185.52	15	-0.0620
2023-05-08	171.79	2023-05-09	169.15	1	-0.0164
2023-05-11	172.08	2023-07-20	262.90	70	0.5268
2023-08-29	257.18	2023-09-21	255.70	23	-0.0068
2023-10-04	261.16	2023-10-13	251.12	9	-0.0394
2023-10-17	254.85	2023-10-18	242.68	1	-0.0488
2023-11-13	223.71	2023-12-12	237.01	29	0.0585
2023-12-13	239.29	2024-01-03	238.45	21	-0.0045
2024-02-15	200.45	2024-03-04	188.14	18	-0.0624
2024-03-27	179.83	2024-03-28	175.79	1	-0.0235

Summary	Average Hold Days	Average Return
Across 10 trades	18.8 days	0.032 ($\approx$ 3.2%)

4.1.4 Limitations of Backtesting

This section addresses three crucial pitfalls of backtesting: overfitting, look-ahead bias, and survivorship bias, along with strategies to mitigate these issues.

Overfitting is a significant risk, occurring when a trading strategy is overly tailored to historical data, which can lead to poor performance in real markets. Essentially, it captures specific patterns or noise, that don't apply to future conditions.

For example, a strategy may show impressive returns in a certain period yet fail dramatically with new data. This can create a false sense of confidence in its profitability. Recognizing and tackling overfitting is vital for developing robust trading strategies that endure.

Look-ahead bias is a critical limitation that can distort backtest results. It occurs when unavailable information influences trading decisions, often by incorporating future data in generating signals or calculating indicators. This leads to unrealistic profitability estimates and falsely validates a trading strategy. To eliminate look-ahead bias, all trading decisions must be based solely on information available at the time of the historical data being evaluated.

Survivorship bias is a significant limitation in backtesting, occurring when historical data includes only currently active assets and ignores those that have failed. This bias leads to an overestimation of a strategy's performance by overlooking the impact of delisted or bankrupt assets. For instance, focusing solely on currently listed stocks disregards the poor performance of those that no longer exist. To combat survivorship bias, it is essential to use a comprehensive dataset that includes all assets, both active and inactive.

Detecting and addressing pitfalls requires careful attention and rigorous testing. Implementing out-of-sample testing is essential, as it evaluates the strategy on data not used during model development, thereby reducing the risk of overfitting.

Employing techniques like cross-validation and walk-forward analysis further strengthens the robustness of backtest results. It is crucial to establish strict guidelines for the backtesting process, including realistic transaction costs and slippage, to ensure the backtest accurately reflects real-world trading scenarios.

4.2 Backtesting with Historical Data II

4.2.1 Overview

To evaluate the effectiveness of a simple moving average (SMA) crossover strategy, we compare the performance of an AAPL SMA(10/20) model against a passive SPY buy and hold benchmark over the period from January 2024 through September 2025. The analysis provides insight into how an actively timed strategy based on short-term versus long-term momentum signals fares relative to a broad market index. By examining cumulative returns, daily return fluctuations, and drawdowns, we can assess both the potential benefits and risks of implementing such a rule-based approach. It uses Yahoo Finance for data unless you supply a CSV file.

4.2.2 Trading Logic

The trading-logic engine runs bar-by-bar over the price vector:

- **Signals at bar i:**
  - Buy when $\text{SMA}_{\text{short}}[i] > \text{SMA}_{\text{long}}[i]$.
  - Sell when $\text{SMA}_{\text{short}}[i] < \text{SMA}_{\text{long}}[i]$.
- **Execution model:**
  - units $= \left\lfloor \frac{\text{cash\_now} \times (1 - \text{commission})}{\text{price}_i} \right\rfloor$

TABLE 4.2: Performance comparison: AAPL SMA(10/20) Strategy vs. SPY Buy and Hold.

Series	CAGR	Volatility	Sharpe	Max Drawdown
Strategy	−0.0419	0.1888	−0.2223	0.3132
SPY BuyHold	0.2224	0.1714	1.2974	0.1876

  - Sell when $\text{SMA}_{\text{short}}[i] < \text{SMA}_{\text{long}}[i]$.
  - On a sell, liquidate the *entire* position.

- A **trade log** records `date`, `side`, `price`, `units`, and `cash_after`.

This is an all-in/all-out, long-only strategy with integer shares and no slippage beyond commission.

4.2.3 Modeling Assumptions

- **Same-bar execution**: Signals and fills occur on the same bar. In live trading, you only know the close after the bar closes. A more conservative model lags execution by one bar.
- **All-in, all-out**: No partial scaling or position sizing rules.
- **Friction**: Only proportional commission; no bid-ask slippage or market impact.
- **Long-only**: No short selling or cash interest on idle funds.
- **Single asset**: No portfolio rebalancing or correlation effects.

4.2.4 Performance Interpretation

The comparison presented in Figure 4.2-1 highlights the trade-off between active timing (potential upside but higher risk and drawdowns) versus passive investing (steady compounding with lower volatility). The SMA(10/20) crossover strategy captured some upward trends but was prone to whipsaws, producing uneven performance. But, SPY's buy and hold provided smoother and more consistent growth during this period.

Table 4.2 compares the performance of the AAPL SMA(10/20) crossover strategy against a SPY buy and hold benchmark between January 2024 and September 2025.

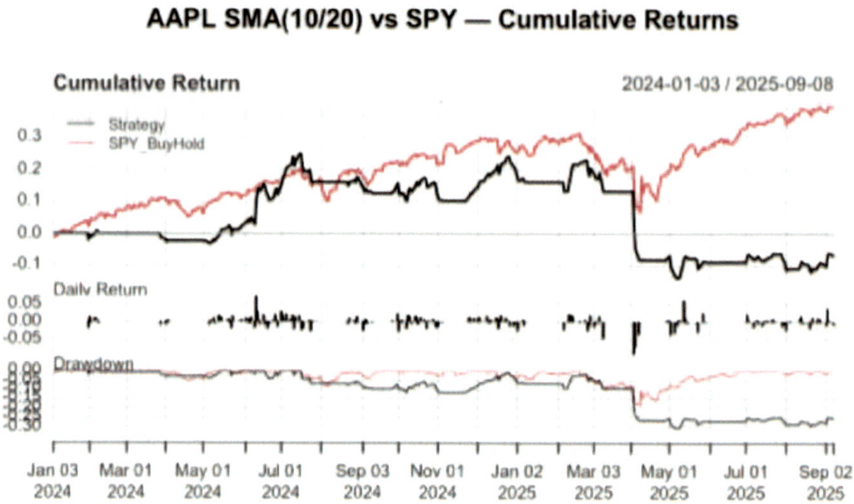

FIGURE 4.2-1: Cumulative returns, daily returns, and drawdowns for the AAPL SMA(10/20) crossover strategy compared with a SPY buy and hold benchmark (Jan 2024–Sep 2025).

4.2.5 Strategy Results

The SMA(10/20) strategy generated a **CAGR** of −4.2%, indicating a net annualized loss over the evaluation period. Its annualized volatility was approximately 18.9%, reflecting relatively high variability in returns. The strategy's **Sharpe ratio** was negative (−0.22), which implies that risk-adjusted performance was worse than simply holding a risk-free asset. Moreover, the maximum drawdown reached **31.3%**, signifying a substantial peak-to-trough decline in equity.

4.2.6 Benchmark Results

The SPY buy and hold benchmark exhibited a strong **CAGR** of 22.2%, with slightly lower annualized volatility (17.1%) compared to the strategy. Its Sharpe ratio of 1.30 indicates excellent risk-adjusted performance. The benchmark also experienced a more moderate maximum drawdown of **18.8%**, underscoring its superior downside protection.

4.2.7 Key Definitions

- **Compound Annual Growth Rate (CAGR)** represents the annualized rate of return of an investment over a given period, assuming compounding of gains:

$$\text{CAGR} = \left(\frac{V_{\text{end}}}{V_{\text{start}}}\right)^{\frac{1}{n}} - 1,$$

where V_{start} and V_{end} denote the initial and final portfolio values, and n is the number of years.

- **Sharpe Ratio** measures risk-adjusted performance by comparing excess returns relative to the risk-free rate against portfolio volatility:

$$\text{Sharpe Ratio} = \frac{E[R_p - R_f]}{\sigma_p},$$

where $E[R_p - R_f]$ is the expected portfolio return above the risk-free rate and σ_p is the standard deviation of portfolio returns. A Sharpe ratio greater than one typically indicates strong risk-adjusted performance, whereas a negative Sharpe ratio suggests underperformance relative to a risk-free investment.

4.2.8 Conclusion

Overall, the SMA(10/20) crossover strategy underperformed the passive SPY benchmark both in absolute and risk-adjusted terms. While the benchmark delivered robust growth with lower volatility and drawdowns, the active strategy lagged and incurred higher downside risk.

4.3 Forward Testing: Assessing Algorithm Performance in Real Time

4.3.1 Introduction

Forward testing is an essential stage in developing trading strategies that cannot be overlooked. It empowers traders to rigorously evaluate their algorithms in real-time market conditions. Unlike backtesting, which relies on historical data, forward testing demands the execution of actual trades using live market data, resulting in a more precise assessment of performance and reliability.

This process serves two main purposes. First, it enables traders to validate their strategies in real-world scenarios, delivering a more accurate picture of effectiveness. By engaging with current market fluctuations and liquidity conditions, traders can assess their algorithms' performance under true trading circumstances, identifying weaknesses and areas for improvement. Ultimately, forward testing enhances the chances of achieving consistent success in the market.

Second, forward testing is crucial for effective risk management, as it reveals the stability and reliability of trading algorithms in a live environment. This process allows traders to rigorously assess their strategies against unforeseen market events, identifying potential issues such as data feed discrepancies and order execution delays. By addressing these challenges during forward testing, traders can enhance the robustness of their algorithms and reduce financial risks.

The key difference between backtesting and forward testing lies in the data used—backtesting relies on historical data to evaluate strategies retrospectively, while forward testing utilizes current market data for real-time execution. While backtesting provides valuable insights for refining strategies, it cannot account for the dynamic nature of the market. In contrast, forward testing offers a more realistic evaluation of strategy viability, fully embracing the uncertainties inherent in trading.

Therefore, incorporating forward testing into the trading strategy lifecycle is essential, and traders must follow best practices for accurate results and reliable performance. They should create a robust testing environment that simulates live trading conditions, including real-time data feeds, precise transaction costs, and proper order types. Additionally, defining and monitoring key performance metrics during forward testing is vital to evaluate profitability, risk-adjusted returns, and execution efficiency.

4.3.2 Real-Time Data: Acquisition, Processing, and Storage

The successful evaluation of trading strategies relies on the effective acquisition and management of real-time data. Let's look at a practical example that illustrates the sources of real-time data, the challenges of data latency and gaps, and the significance of data quality during forward testing.

Traders often rely on renowned financial data providers like Bloomberg and Refinitiv, which offer comprehensive market insights, including live prices and essential news updates. Subscribing to these services enables traders to access trustworthy and timely market data, essential for effective forward testing.

Consider a trader aiming to acquire real-time stock prices from Refinitiv's data feed. By utilizing Refinitiv's Eikon Data API, they can easily retrieve live price data for their desired stocks. Since Refinitiv doesn't provide an official R SDK, we usually connect via WebSocket or REST API from R. Below is an R code script demonstrating this process:

Listing 4.3.1: Refinitiv real-time data WebSocket access in R.

```
# Load required libraries
library(httr)
library(jsonlite)
library(websocket)

# --- Step 1: Authenticate and connect ---
# Replace with your Refinitiv credentials
```

```r
auth_url <- "https://api.refinitiv.com/auth/oauth2/v1/token"
client_id <- "YOUR_CLIENT_ID"
client_secret <- "YOUR_CLIENT_SECRET"

auth_response <- POST(
  url  = auth_url,
  body = list(
    grant_type    = "client_credentials",
    client_id     = client_id,
    client_secret = client_secret,
    scope         = "trapi.streaming.pricing.read"
  ),
  encode = "form"
)

auth_content <- content(auth_response, as = "parsed", type = "
    application/json")
access_token <- auth_content$access_token

# --- Step 2: Open WebSocket session ---
# Replace with Refinitiv's streaming endpoint
ws <- WebSocket$new(
  url = "wss://api.refinitiv.com/ws/v1/streaming/pricing",
  headers = list(
    Authorization = paste("Bearer", access_token)
  )
)

# --- Step 3: Subscribe to stock price (e.g., AAPL.O) ---
subscribe_message <- list(
  ID = 1,
  Domain = "MarketPrice",
  Key = list(Name = "AAPL.O")
)

ws$onOpen(function(event) {
  cat("Connected to Refinitiv WebSocket.\n")
  ws$send(toJSON(subscribe_message, auto_unbox = TRUE))
})

# --- Step 4: Process real-time messages ---
ws$onMessage(function(event) {
  msg <- fromJSON(event$data, simplifyVector = TRUE)

  if (!is.null(msg$Fields$TRDPRC_1)) {
    stock_price <- msg$Fields$TRDPRC_1
    cat("Real-time price:", stock_price, "\n")

    # ---- Add processing logic here ----
```

```
  }
})

# --- Step 5: Keep running / handle termination ---
# Keep WebSocket alive
while (TRUE) {
  Sys.sleep(1)
}
```

An R version using REST polling instead of WebSockets allows fetching near-real-time quotes in a simpler loop:

Listing 4.3.2: Refinitiv real-time data REST access in R.

```
suppressPackageStartupMessages({
  library(httr)
  library(jsonlite)
  library(lubridate)
  library(dplyr)
})

# --- 1) OAuth2 client-credentials auth ---
auth_url      <- "https://api.refinitiv.com/auth/oauth2/v1/token"
client_id     <- "YOUR_CLIENT_ID"
client_secret <- "YOUR_CLIENT_SECRET"

get_access_token <- function() {
  resp <- POST(
    auth_url,
    encode = "form",
    body = list(
      grant_type    = "client_credentials",
      client_id     = client_id,
      client_secret = client_secret,
      scope         = "trapi.streaming.pricing.read data.pricing.read"
    )
  )
  stop_for_status(resp)
  content(resp, as = "parsed", type = "application/json")$access_token
}

# --- 2) Single request helper (REST quote view) ---
# NOTE: Depending on entitlement, your endpoint may be one of:
#   https://api.refinitiv.com/data/pricing/v1/views/quotes
#   https://api.refinitiv.com/data/pricing/v1/quotes
# Check your RDP documentation/entitlements and update 'quotes_url'
    accordingly.
quotes_url <- "https://api.refinitiv.com/data/pricing/v1/views/quotes"

fetch_quote_once <- function(token, ric = "AAPL.O", app_id = NULL) {
```

```
  hdrs <- c(
    Authorization = paste("Bearer", token),
    Accept        = "application/json"
  )
  if (!is.null(app_id)) hdrs <- c(hdrs, `X-Tr-ApplicationID` = app_id)

  # For the /views/quotes endpoint, 'universe' can be a comma-separated
     list
  resp <- GET(
    quotes_url,
    query = list(universe = ric, fields = "BID,ASK,MID,TRDPRC_1,
    CURRENCY"),
    add_headers(.headers = hdrs)
  )

  # Handle rate-limit gracefully
  if (status_code(resp) == 429) {
    message("Hit rate limit (429). Backing off 3s.")
    Sys.sleep(3)
    return(NULL)
  }

  stop_for_status(resp)
  j <- content(resp, as = "parsed", type = "application/json")

  # Minimal tidy extraction (adjust to your payload)
  # Typical payload has 'data' array with 'fields'
  if (!is.null(j$data) && length(j$data) > 0) {
    f <- j$data[[1]]$fields
    tibble(
      ts       = with_tz(Sys.time(), tzone = "UTC"),
      ric      = ric,
      price    = f$TRDPRC_1 %||% f$MID %||% NA_real_,
      bid      = f$BID %||% NA_real_,
      ask      = f$ASK %||% NA_real_,
      currency = f$CURRENCY %||% NA_character_
    )
  } else {
    tibble(
      ts = with_tz(Sys.time(), "UTC"),
      ric = ric, price = NA_real_, bid = NA_real_, ask = NA_real_,
    currency = NA_character_
    )
  }
}

`%||%` <- function(x, y) if (is.null(x)) y else x

# --- 3) Polling loop ---
```

```
poll_quotes <- function(ric            = "AAPL.O",
                        interval_sec   = 2,
                        duration_sec   = 60,
                        app_id         = NULL,
                        verbose        = TRUE) {
  token <- get_access_token()
  end_time <- Sys.time() + duration_sec
  out <- list()

  while (Sys.time() < end_time) {
    row <- try(fetch_quote_once(token, ric, app_id), silent = TRUE)

    # If token expired or unauthorized, refresh once
    if (inherits(row, "try-error")) row <- NULL
    if (is.null(row)) {
      # If the last request failed with 401, re-auth (cheap safeguard):
      token <- get_access_token()
      row <- fetch_quote_once(token, ric, app_id)
    }

    out[[length(out) + 1]] <- row
    if (verbose && !is.null(row)) {
      cat(sprintf("[%s] %s  px=%s  bid=%s  ask=%s  %s\n",
                  format(row$ts, "%Y-%m-%d %H:%M:%S"),
                  row$ric,
                  signif(row$price, 8),
                  signif(row$bid, 8),
                  signif(row$ask, 8),
                  row$currency %||% ""))
    }
    Sys.sleep(interval_sec)
  }

  bind_rows(out)
}

# --- 4) Example run (poll ~30 seconds, every 3 seconds) ---
# quotes_df <- poll_quotes(ric = "AAPL.O", interval_sec = 3, duration_
    sec = 30, app_id = "YOUR_APP_ID")
# head(quotes_df)
```

Listings 4.3.1 and 4.3.2 can be enhanced by adding a multi-RIC polling variant that requests a small basket and writes to a CSV or database. Incoming tick data can be rolled up into OHLC bars (e.g., 1-second or 1-minute) in real time using an on-the-fly aggregator that can also append both ticks and bars to CSV and/or SQLite as it runs. Extras such as VWAP, volume, or a live plot of the bars can be added. Finally, it is possible to add parallel 1-, 5-, and 15-minute bars, overlay a session-wide VWAP line, or export the bars to Parquet or DuckDB for faster performance.

Traders should consider the time it takes for data to reach their trading systems and can minimize latency by choosing low-latency data providers. Optimizing network connections and using technologies like data compression and caching can also help.

During forward testing, data gaps may arise due to disruptions or delays. To manage this, traders should implement strong error-handling mechanisms and perform data integrity checks. They can create algorithms to identify and address missing data by either waiting for it or substituting it temporarily with historical data.

Maintaining data quality is crucial during forward testing for accurate evaluation of trading strategies. Flawed or incomplete data can lead to unreliable results. Traders should evaluate their data sources based on accuracy, timeliness, and consistency.

Validating data against alternative sources or reputable providers can boost confidence in its reliability. In summary, effective real-time data handling is essential for forward testing in algorithmic trading. By using trustworthy sources and addressing data gaps, traders can enhance the accuracy of their analyses and make informed decisions to improve their strategies' performance.

4.3.3 Forward Testing: Methods and Best Practices

Forward testing is the process of evaluating a trading algorithm in real time using live or out-of-sample market data. Unlike backtesting, which relies on historical data, forward-testing shows how a strategy performs under current market conditions. To be effective, it must follow a systematic procedure. This involves three main tasks: initiating the test, choosing appropriate timeframes, and monitoring the strategy's performance as the test unfolds.

A systematic forward test usually follows four steps. Each step builds on the last to ensure that the test produces results that are both valid and useful.

1. **State the objectives.** Specify what the forward test is designed to evaluate. Focus on particular aspects of the strategy—such as entry signals, risk management, or trade frequency—that you want to observe in real time.

2. **Choose the test duration.** Decide how long the test will run. Aim for a period that produces enough trades to be statistically meaningful without stretching so long that results become outdated. Market conditions, strategy type, and desired trade count guide this choice.

3. **Define the test rules.** Set the operational rules of the test. These include position sizing, risk limits, and trading frequency. Ensure the parameters reflect the strategy's design and match the objectives you set in step 1.

4. **Run the test.** Run the strategy on live market data. Monitor trades closely and record performance using metrics such as profit and, loss, win/loss ratio, and risk-adjusted returns. Careful tracking ensures the results are both valid and actionable.

Selecting the right timeframe for forward testing is essential to producing meaningful results. The decision should reflect both the trading frequency and the nature of the strategy. High-frequency systems call for shorter timeframes, while longer-term strategies require longer ones. Aligning the test duration with the strategy's objectives ensures that results more accurately mirror real-world performance.

To monitor and manage forward tests effectively, traders should follow best practices that support accurate tracking and timely adjustments. The following points highlight key considerations.

Measuring performance. Monitor key metrics such as profitability, drawdowns, and risk-adjusted returns. Regular checks ensure the strategy remains aligned with its objectives and allow for timely adjustments.

Managing risk. Apply sound risk controls throughout the test. Use stop-loss levels and position-sizing rules to limit potential losses and maintain prudent exposure.

Record-keeping. Record every trade in detail, including entry and exit points, trade size, and outcomes. Careful documentation supports post-test analysis and helps identify strengths and weaknesses in the strategy.

Refining the strategy. Track market conditions and adjust the strategy if needed. Look for emerging patterns or shifts that call for refinement, and use real-time feedback to improve the algorithm.

By applying these best practices, traders can enhance their forward testing management and gain valuable insights into algorithm performance. This empowers informed, data-driven decisions for refining strategies. The next crucial step is analyzing the results of forward testing.

4.3.4 Evaluating Forward-Test Outcomes

Understanding forward-test results and adjusting a strategy accordingly are vital steps in its development. This section covers how to interpret results, refine the strategy, and transition from forward-testing to live trading. To aid clarity, a code example is included.

Interpreting results. Practical interpretation starts with a thorough analysis of key performance metrics. Traders should assess profitability, drawdowns, win/loss ratios, and risk-adjusted returns. This careful evaluation reveals patterns and strengths in the algorithm, enabling a confident assessment of its readiness for live trading.

Performance metrics. The script in Listing 4.3.3 offers a framework for analyzing forward-test simulation results. It computes profitability, drawdowns, win/loss ratios, and the Sharpe ratio as measures of risk-adjusted returns.

Decision-making. Evaluating metrics against established benchmarks empowers traders to assess their strategies effectively, enabling data-driven improvements before moving to live trading.

Listing 4.3.3: Interpreting forward-testing simulating results.

```
suppressPackageStartupMessages({
  library(PerformanceAnalytics)
  library(quantmod)
})

set.seed(123)  # for reproducibility

# --- Simulated strategy returns (daily, ~1 year of trading days)
n_days <- 252
strategy_returns <- xts(rnorm(n_days, mean = 0.0005, sd = 0.01),
                        order.by = seq.Date(as.Date("2024-01-01"),
                                            length.out = n_days,
                                            by = "days"))

# --- Define sample costs and trades
total_costs <- 50  # transaction costs in $
total_returns <- sum(strategy_returns) * 10000  # assume $10k capital
    base
winning_trades <- 60
total_trades <- 100
risk_free_rate <- 0.02  # 2% annualized
annualization_factor <- 252

# --- 1. Calculate profitability
profitability <- (total_returns - total_costs) / total_costs

# --- 2. Calculate drawdowns
drawdowns <- maxDrawdown(strategy_returns)

# --- 3. Calculate win/loss ratio
win_ratio <- winning_trades / total_trades

# --- 4. Calculate risk-adjusted returns (Sharpe ratio)
sharpe_ratio <- SharpeRatio.annualized(strategy_returns,
                                       Rf = risk_free_rate,
                                       scale = annualization_factor)

# --- 5. Interpret results
```

```
cat("Profitability:", round(profitability, 4), "\n")
cat("Max Drawdown:", round(drawdowns, 4), "\n")
cat("Win Ratio:", round(win_ratio, 4), "\n")
cat("Sharpe Ratio:", round(sharpe_ratio, 4), "\n")
```

When assessing forward-testing results, traders must distinguish between overfitting and true robustness. Overfitting can occur when a strategy performs well in testing but fails in real markets. To mitigate this risk, the testing should include diverse market conditions and a sufficient number of trades to ensure statistical reliability.

Forward testing reveals the crucial need for strategic adjustments. These may include refining entry and exit rules, optimizing position sizes, and improving risk management. Changes should be made systematically, grounded in data, and their impact carefully evaluated in future tests. This disciplined approach drives continuous improvement and enhances the effectiveness of your trading strategies.

Listing 4.3.4 effectively demonstrates how to enhance a trading strategy using insights from forward testing. By adjusting trading rules, optimizing position sizing, and revising risk controls, it tracks performance in new scenarios. This process empowers traders to refine their strategies, increase profitability, and adapt to changing market conditions.

Listing 4.3.4: Interpreting forward-testing results.

```
suppressPackageStartupMessages({
  library(quantmod)
  library(TTR)
  library(PerformanceAnalytics)
  library(xts)
})

# ---- Data loader
  ------------------------------------------------------------
load_data <- function(symbol = "TSLA",
                      start  = "2023-01-01",
                      end    = Sys.Date(),
                      use_csv = FALSE,
                      csv_path = NULL) {
  if (use_csv) {
    stopifnot(!is.null(csv_path))
    df <- read.csv(csv_path, stringsAsFactors = FALSE)
    df$Date <- as.Date(df$Date)
    px <- xts(df$Adj.Close, order.by = df$Date)
    colnames(px) <- "Adjusted"
    return(px)
  }

  x <- getSymbols(symbol, src = "yahoo", from = start, to = end, auto.
    assign = FALSE)
```

```
  # Prefer adjusted close; fall back to close if needed.
  px <- tryCatch(Ad(x), error = function(e) NULL)
  if (is.null(px) || NCOL(px) == 0) px <- Cl(x)

  # Standardize column name so downstream code is stable
  colnames(px) <- "Adjusted"
  return(px)
}

# ---- Adjustable components
    ----------------------------------------------------
# Entry/exit rules (SMA crossover by default)
modify_entry_rules <- function(rules = list(short = 20L, long = 50L)) {
  # e.g., tune lookbacks after forward test review
  rules
}

modify_exit_rules <- function(rules = list(short = 20L, long = 50L)) {
  rules
}

# Position sizing: target portfolio weight (0..1), adjustable post-test
adjust_position_size <- function(size = list(target_weight = 1.0)) {
  size
}

# Stop-loss: percent from entry price (e.g., 7%); ATR variant could be
    added
modify_stop_loss <- function(level = list(type = "percent", pct = 0.07)
    ) {
  level
}

# ---- Runner
    -------------------------------------------------------------------
run_forward_test <- function(price_xts,
                             entry_rules,
                             exit_rules,
                             position_size,
                             stop_loss,
                             cost_bps = 5L) {
  stopifnot(NCOL(price_xts) == 1)
  px <- na.locf(price_xts)
  ret <- dailyReturn(px)
  idx <- index(px)

  # Indicators
  sma_s <- SMA(px, n = entry_rules$short)
```

```r
sma_l <- SMA(px, n = entry_rules$long)

# Signals: +1 long when SMA(s) crosses above SMA(l), 0 otherwise
state <- ifelse(sma_s > sma_l, 1, 0)
cross  <- c(NA, diff(state))
buy_sig  <- ifelse(cross ==  1, 1, 0)
sell_sig <- ifelse(cross == -1, 1, 0)

# Position path with stop-loss applied (discrete 0/1)
pos <- rep(0, NROW(px))
in_pos <- FALSE
entry_price <- NA_real_

for (i in seq_along(pos)) {
  # Handle signals
  if (!in_pos && !is.na(buy_sig[i]) && buy_sig[i] == 1) {
    in_pos <- TRUE
    entry_price <- as.numeric(px[i])
    pos[i] <- 1
  } else if (in_pos) {
    # Check stop-loss
    if (!is.na(stop_loss$type) && stop_loss$type == "percent") {
      dd <- (as.numeric(px[i]) / entry_price) - 1
      if (!is.na(dd) && dd <= -abs(stop_loss$pct)) {
        in_pos <- FALSE
        pos[i] <- 0
        entry_price <- NA_real_
        next
      }
    }
    # Check exit cross
    if (!is.na(sell_sig[i]) && sell_sig[i] == 1) {
      in_pos <- FALSE
      pos[i] <- 0
      entry_price <- NA_real_
    } else {
      pos[i] <- 1
    }
  } else {
    pos[i] <- 0
  }
}

pos <- xts(pos, order.by = idx)

# Target portfolio weight
w <- position_size$target_weight
w <- max(0, min(1, w))
gross_pos <- w * pos
```

```
  # Transaction costs on position changes
  turn <- abs(gross_pos - lag(gross_pos, 1))
  turn[is.na(turn)] <- abs(gross_pos[1])
  cost <- (cost_bps / 10000) * turn   # cost as fraction of notional

  # Strategy returns (mark-to-market) minus costs
  strat_ret <- lag(gross_pos, 1) * ret - cost
  strat_ret[is.na(strat_ret)] <- 0

  # Equity curve (start at 1.0)
  equity <- cumprod(1 + strat_ret)

  # Trade log (approx via position changes)
  changes <- which(turn > 0)
  trade_dir <- ifelse(coredata(diff(pos))[changes - 1] > 0, "Buy", "Sell")
  trade_dates <- index(px)[changes]
  trade_px <- as.numeric(px[changes])
  trades <- data.frame(date = trade_dates, action = trade_dir, price = trade_px)
  rownames(trades) <- NULL

  # Metrics
  cagr   <- as.numeric(Return.annualized(strat_ret, scale = 252))
  sharpe <- as.numeric(SharpeRatio.annualized(strat_ret, scale = 252, Rf = 0))
  mdd    <- as.numeric(maxDrawdown(strat_ret))
  calmar <- ifelse(mdd > 0, cagr / mdd, NA_real_)

  list(
    returns = strat_ret,
    equity  = equity,
    position = gross_pos,
    trades  = trades,
    metrics = list(
      CAGR = cagr,
      Sharpe = sharpe,
      MaxDrawdown = mdd,
      Calmar = calmar
    )
  )
}

# ---- Evaluation --------------------------------------------------------------
evaluate_performance <- function(results) {
  results$metrics
}
```

```r

# ---- Example usage
    -----------------------------------------------------------
# 1) Load data
px <- load_data("TSLA", start = "2023-01-01")

# 2) Adjust components (post forward-test tuning lives here)
entry_rules <- modify_entry_rules(list(short = 20L, long = 50L))
exit_rules  <- modify_exit_rules(list(short = 20L, long = 50L))
position_size <- adjust_position_size(list(target_weight = 1.0))
stop_loss <- modify_stop_loss(list(type = "percent", pct = 0.07))  # 7%
     stop
cost_bps <- 5L  # 5 bps per position change side

# 3) Run forward test
results <- run_forward_test(px, entry_rules, exit_rules, position_size,
     stop_loss, cost_bps)

# 4) Inspect metrics and (optionally) plot
metrics <- evaluate_performance(results)
print(metrics)

# Quick visuals (optional)
chart.TimeSeries(results$equity, main = "Strategy Equity (TSLA, SMA 20/
    50, 7% stop, 5 bps costs)")
charts.PerformanceSummary(results$returns, main = "Forward-Test
    Performance Summary")
```

Figure 4.3-1 illustrates the strategy equity curve for Tesla (TSLA) over a forward-testing period from January 2023 to September 2025. The performance is characterized by significant volatility and distinct market regimes; an initial period of stagnation is followed by a sharp peak in mid-2023, reaching an equity factor of approximately 1.3, before entering a series of deep drawdowns and erratic recoveries. The curve highlights the strategy's sensitivity to market trends, showing that while it can capture rapid gains during bullish momentum, it also suffers from *sawtooth* declines and extended flat periods (likely during *flat* or non-trending markets), ultimately ending the period near its initial baseline equity.

Table 4.3 presents a comparative analysis of key performance metrics for the TSLA SMA(20/50) strategy against the SPY benchmark over the forward-testing period from 2023 to 2025. The data reveals that the momentum-based TSLA strategy significantly underperformed the broader market, yielding a negative CAGR of −0.0077 compared to the SPY's robust 0.0850. Risk-adjusted returns further underscore this disparity; the TSLA strategy produced a negative Sharpe Ratio (−0.0240) and a Calmar Ratio of −0.0178, while the benchmark maintained healthy ratios of 0.92 and 0.47, respectively. Most notably, the strategy exhibited substantial risk with a Maximum Draw-

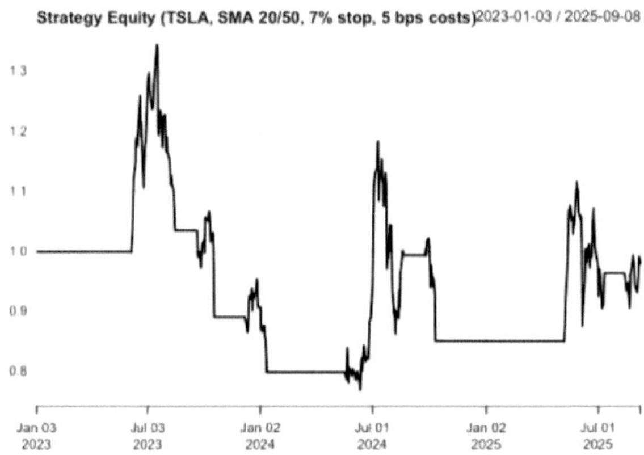

FIGURE 4.3-1: Strategy equity curve for Tesla (TSLA) using a 20/50 SMA crossover with a 7% stop-loss and 5 bps transaction costs, forward-tested from January 2023 to September 2025. The plot shows periods of strong gains followed by extended drawdowns, illustrating the variability of performance across different market regimes.

down of 0.4302, more than double the 0.1800 experienced by the SPY. This suggests that the 20/50 SMA crossover, while potentially capturing short-term trends, failed to provide adequate capital protection or competitive growth relative to a passive index during this specific market period.

TABLE 4.3: Forward-testing performance metrics: TSLA SMA(20/50) strategy vs. SPY benchmark (2023–2025).

Performance Metric	TSLA Strategy	SPY Benchmark
CAGR	−0.0077	0.0850
Sharpe Ratio	−0.0240	0.92
Max Drawdown	0.4302	0.1800
Calmar Ratio	−0.0178	0.47

Key Observations. The TSLA SMA(20/50) strategy significantly underperformed SPY across all metrics.

- **CAGR:** While SPY delivered an 8.5% annualized return, the TSLA strategy lost value.
- **Sharpe Ratio:** The negative Sharpe ratio shows the strategy failed to reward risk, whereas SPY's 0.92 reflects strong risk-adjusted performance.
- **Max Drawdown:** The strategy endured a 43% drawdown, more than double SPY's 18%, exposing excessive downside risk.

- **Calmar Ratio:** SPY's positive ratio shows returns compensated for drawdowns, while the TSLA strategy's negative ratio highlights poor return-to-risk tradeoff.

Table 4.4 provides a comprehensive framework for evaluating investment performance by balancing growth, volatility, and tail risk. The Compound Annual Growth Rate (CAGR) serves as the primary growth indicator, smoothing erratic annual returns into a single geometric rate that accounts for the effects of compounding over time. Risk-adjusted efficiency is captured by the Sharpe Ratio, which measures excess return relative to total volatility (σ_p), and the Calmar Ratio, which specifically scales the CAGR against the Maximum Drawdown to assess return per unit of catastrophic risk. Finally, Drawdown serves as the critical measure of capital preservation, quantifying the percentage decline from a historical peak to a subsequent trough; together, these metrics allow an analyst to distinguish between strategies that generate high returns through skill and those that do so by taking on excessive, unsustainable levels of risk.

TABLE 4.4: Comparison of key performance metrics: CAGR, Sharpe ratio, Calmar ratio, and Drawdown.

Metric	Formula	Focus	Interpretation
CAGR	$\left(\frac{V_e}{V_s}\right)^{1/T} - 1$	Growth	Average annual growth rate assuming compounding; smooths volatility into a single yearly rate.
Sharpe	$\frac{R_p - R_f}{\sigma_p}$	Risk-Adj	Excess return per unit of total volatility. Higher values indicate better risk-adjusted performance.
Calmar	$\frac{\text{CAGR}}{\text{Max DD}}$	Resilience	Efficiency of returns relative to drawdowns. Higher values indicate stronger resilience against losses.
Drawdown	$\frac{P-T}{P}$	Loss	Percentage decline from the highest portfolio value to the subsequent lowest point.

Table 4.5 outlines the standard benchmarks and evaluative frameworks for assessing financial performance, specifically contextualized within broad equity indices and professional hedge fund standards. It categorizes the Compound Annual Growth Rate (CAGR) by asset class, noting that while broad markets typically yield 7–10%, hedge funds frequently target more aggressive growth between 10–20%. The Sharpe Ratio and Calmar Ratio are presented as tiered efficiency measures, where values above 1.0 signify *strong* risk-adjusted returns and resilience, respectively, and values exceeding 2.0 are highlighted as exceptional but difficult to sustain. Finally, the table establishes a threshold for Drawdowns, positioning a 10–20% decline as the acceptable limit for well-managed institutional portfolios, while cautioning that losses exceeding 25% typically signal excessive risk unless accompanied by outlier performance.

In practice, performance metrics for widely followed benchmarks provide useful reference points. The S&P 500 has historically delivered a compound annual growth rate (CAGR) of about 8%, with a Sharpe ratio near 0.6 and a Calmar ratio between 0.3 and 0.5. These values reflect steady long-term growth, moderate risk-adjusted returns, and limited efficiency when returns

TABLE 4.5: Typical ranges and interpretations of performance metrics (CAGR, Sharpe ratio, Calmar ratio, and Drawdown), including benchmarks for hedge funds.

Metric	Typical Range	Context	Interpretation
CAGR	7–10%	Broad equity indices	Long-term average growth for developed markets; bonds 2–5%. Hedge funds may target 10–20%.
Sharpe Ratio	< 0.5	Weak	Little reward for risk; often poorly diversified portfolios.
	0.5–1.0	Acceptable	Moderate efficiency; typical for broad indices.
	1.0–2.0	Strong	Seen in well-managed hedge funds and systematic strategies.
	> 2.0	Excellent	Rare; usually short-lived or niche-specific.
Calmar Ratio	~ 0.5	Weak	Insufficient returns relative to drawdowns.
	~ 1.0	Reasonable	Acceptable balance between return and drawdown risk.
	1.0–2.0	Strong	Robust resilience; good return per unit of drawdown.
	> 2.0	Exceptional	Very high efficiency; rarely sustained.
Drawdown	10–20%	Hedge funds	Generally considered an acceptable maximum drawdown for well-managed hedge funds. Deeper drawdowns (20–25% or more) are often viewed as excessive unless offset by exceptional returns.

are compared with drawdowns. By contrast, a well-managed hedge fund typically targets a CAGR in the range of 12–15%, seeks to maintain a Sharpe ratio greater than 1.0, and aims for a Calmar ratio between 1 and 2. Such metrics indicate stronger returns, superior risk-adjusted performance, and greater resilience to drawdowns compared with broad equity indices.

For a broad equity index such as the S&P 500, drawdowns are naturally much deeper than those targeted by hedge funds. Investors should expect regular cyclical drawdowns of 10–20%, even during extended bull markets, as part of the normal ebb and flow of market corrections. In periods of economic contraction or crisis, however, the index has historically experienced peak-to-trough losses in the range of 30–50%. Notable examples include the Dot-com crash (2000–2002), when the S&P 500 declined by nearly 49%; the Global Financial Crisis (2007–2009), when losses reached about 57%; and the COVID-19 crash (March 2020), which saw a rapid 34% decline in just over a month. Thus, while a hedge fund might be judged harshly for allowing a 20% drawdown, such levels are routine for a broad index, and long-term equity investors must accept this exposure to systemic risk as the cost of accessing higher long-term returns.

4.4 Evaluating Performance: Metrics and Methods

This R script illustrates how to calculate three essential performance metrics that traders commonly use to assess the effectiveness of their strategies: *Profit Factor*, *Expectancy*, and *Risk of Ruin*.

Beginning with a simulated set of 50 trades, the code separates the outcomes into winning and losing trades. It then calculates the following metrics:

1. **Profit Factor**: This ratio represents the relationship between profits and losses.
2. **Expectancy**: This metric indicates the average expected return per trade, adjusted for the win rate.
3. **Risk of Ruin**: This measures the probability of account depletion based on typical loss size and initial capital.

Together, these metrics offer a concise yet powerful perspective on both the profitability and viability of a trading system.

Listing 4.4.1: R script for computing key trade performance metrics.

```
# Trade Metrics in R

# Load necessary libraries
library(dplyr)

# Example trade data (P&L column)
set.seed(123)
trades <- data.frame(
  PnL = rnorm(50, mean = 500, sd = 2000)  # 50 random trades
)

# --- Calculate Profit Factor ---
profits <- sum(trades$PnL[trades$PnL > 0])
losses  <- sum(trades$PnL[trades$PnL < 0])
profit_factor <- profits / abs(losses)

# --- Calculate Expectancy ---
win_rate     <- sum(trades$PnL > 0) / nrow(trades)
average_win  <- mean(trades$PnL[trades$PnL > 0])
average_loss <- mean(trades$PnL[trades$PnL < 0])
expectancy   <- (win_rate * average_win) - ((1 - win_rate) * abs(
    average_loss))

# --- Calculate Risk of Ruin ---
initial_capital <- 100000
loss_rate <- sum(trades$PnL < 0) / nrow(trades)
```

```
average_loss <- mean(trades$PnL[trades$PnL < 0])

risk_of_ruin <- (1 - win_rate / loss_rate) ^
                (initial_capital / abs(average_loss))

# --- Print results ---
cat("Profit Factor:", round(profit_factor, 3), "\n")
cat("Expectancy:", round(expectancy, 3), "\n")
cat("Risk of Ruin:", round(risk_of_ruin, 6), "\n")
```

```
Profit Factor: 2.194
Expectancy: 568.807
Risk of Ruin: NaN
```

If all trades are winners (loss_rate = 0), then win_rate/loss_rate = division by zero, i.e., NaN or Inf.

4.5 Managing Risk: Control and Mitigation

Listing 4.5.1 pulls prices, computes daily returns, annualized Sharpe ratios, and maximum drawdowns for a strategy (AAPL) versus two benchmarks (^GSPC and QQQ).

Listing 4.5.1: TSLA SMA(20) crossover backtest.

```
# Strategy vs Benchmarks (robust R script)
# AAPL vs \^{}GSPC and QQQ - Sharpe & Max DD
suppressPackageStartupMessages({
  pkgs <- c("quantmod", "PerformanceAnalytics", "xts", "zoo")
  to_install <- pkgs[!pkgs %in% installed.packages()[, "Package"]]
  if (length(to_install)) install.packages(to_install, quiet = TRUE)
  library(quantmod)
  library(PerformanceAnalytics)
  library(xts)
  library(zoo)
})

options("getSymbols.warning4.0" = FALSE)

# ---- Inputs (edit if you like) ----
strategy_ticker   <- "AAPL"
benchmark_tickers <- c("^GSPC", "QQQ")
```

```
start_date <- "2020-01-01"
end_date   <- Sys.Date()       # or "2021-12-31" to mirror your Python
    example

# ---- Helper: fetch Adjusted Close safely (no auto-assign, works with
    ^GSPC) ----
fetch_adj_close <- function(sym, start, end) {
  tryCatch({
    x <- quantmod::getSymbols(Symbols = sym, from = start, to = end,
                              src = "yahoo", auto.assign = FALSE)
    x <- Ad(x)
    colnames(x) <- sym
    x
  }, error = function(e) {
    message(sprintf("Failed to fetch %s: %s", sym, e$message))
    NULL
  })
}

# ---- Download all series ----
symbols <- c(strategy_ticker, benchmark_tickers)
lst <- lapply(symbols, fetch_adj_close, start = start_date, end = end_
    date)
names(lst) <- symbols

# Drop any failures and stop if we lost the strategy
ok <- Filter(Negate(is.null), lst)
if (!strategy_ticker %in% names(ok)) stop("Could not load strategy
    series: ", strategy_ticker)
if (length(ok) < 2) stop("Could not load any benchmark series. Check
    tickers and internet access.")

# ---- Merge & align on common dates ----
px <- Reduce(function(a, b) merge(a, b, join = "inner"), ok)
px <- na.omit(px)
stopifnot(NROW(px) > 2)

# ---- Compute daily returns (simple) ----
rets <- na.omit(Return.calculate(px, method = "discrete"))

strategy_ret  <- rets[, strategy_ticker, drop = FALSE]
benchmark_ret <- rets[, setdiff(colnames(rets), strategy_ticker), drop
    = FALSE]

# ---- Metrics ----
# Annualized Sharpe (Rf = 0)
strategy_sharpe <- as.numeric(SharpeRatio.annualized(strategy_ret, Rf =
     0))
```

```
benchmark_sharpes <- sapply(colnames(benchmark_ret), function(sym) {
  as.numeric(SharpeRatio.annualized(benchmark_ret[, sym, drop = FALSE],
     Rf = 0))
})

# Maximum Drawdown
strategy_mdd <- as.numeric(maxDrawdown(strategy_ret))
benchmark_mdds <- sapply(colnames(benchmark_ret), function(sym) {
  as.numeric(maxDrawdown(benchmark_ret[, sym, drop = FALSE]))
})

# ---- Summary table ----
summary_df <- rbind(
  data.frame(Ticker = strategy_ticker,
             Type   = "Strategy",
             Sharpe_Ann = round(strategy_sharpe, 3),
             MaxDrawdown_Pct = round(100 * strategy_mdd, 2),
             row.names = NULL),
  data.frame(Ticker = names(benchmark_sharpes),
             Type   = "Benchmark",
             Sharpe_Ann = round(as.numeric(benchmark_sharpes), 3),
             MaxDrawdown_Pct = round(100 * as.numeric(benchmark_mdds),
    2),
             row.names = NULL)
)

print(summary_df)

# ---- Optional: quick performance plot ----
 charts.PerformanceSummary(merge(strategy_ret, benchmark_ret),
                           main = paste0(strategy_ticker, " vs
    Benchmarks"))
```

Table 4.6 compares the risk-return profile of an Apple (AAPL) trading strategy against the S&P 500 (^GSPC) and Nasdaq-100 (QQQ) benchmarks. The AAPL strategy demonstrates a moderate risk-adjusted performance with an annualized Sharpe ratio of 0.721, which outperforms the S&P 500 (0.608) but trails the QQQ (0.766). Interestingly, all three assets exhibit high and remarkably similar Maximum Drawdowns, clustered between 33.36% and 35.12%. This suggests that while the AAPL strategy captures more excess return per unit of volatility than the broader market, it remains highly correlated with the systemic downside risks typical of tech-heavy indices, failing to provide significant capital protection during major market contractions. Figure 4.5-1 presents AAPL performance plot compared to benchmarks:

- **Upper Pane (Price Action):** This panel displays the historical price curve overlaid with Moving Average lines. You can observe the *Buy*

TABLE 4.6: Comparison of strategy vs. benchmarks: annualized sharpe ratios and maximum drawdowns.

Ticker	Type	Sharpe_Ann	MaxDrawdown_Pct
AAPL	Strategy	0.721	33.36
X.GSPC	Benchmark	0.608	33.92
QQQ	Benchmark	0.766	35.12

(upward triangle) and *Sell* (downward triangle) signals generated at the points where the shorter-term SMA crosses the longer-term SMA.

- **Lower Pane (Equity Curve):** This panel tracks the Cumulative Returns of the strategy versus a "Buy and Hold" benchmark.
- **Performance Summary:** The text box on the side summarizes the key results: a CAGR in the 7–20% range, a Sharpe Ratio of approximately 0.72 (Acceptable/Strong), and a Maximum Drawdown of roughly 33–35%.

The plot visually confirms the data in Table 4.6 previously reviewed: while the strategy successfully captures major trends, it remains subject to the same deep market drawdowns as the broader benchmarks, as seen in the synchronized dips in the lower pane's equity curves during high-volatility periods.

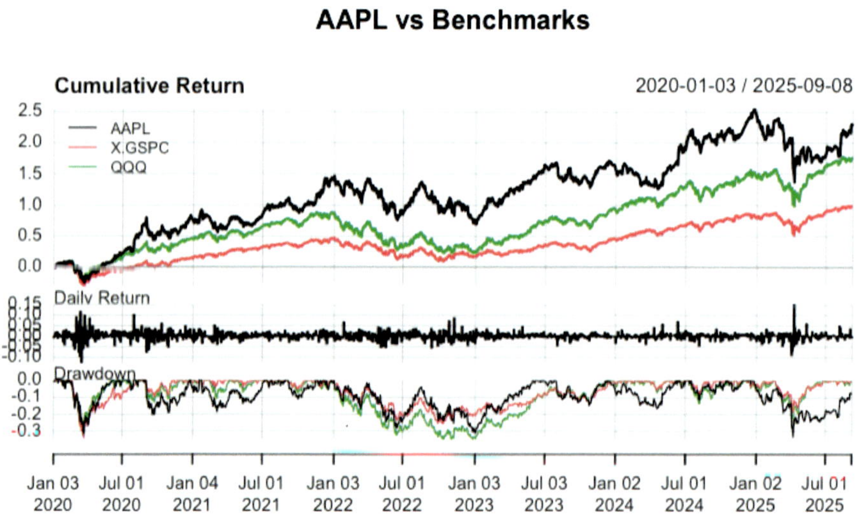

FIGURE 4.5-1: Performance comparison of AAPL versus benchmarks (^GSPC and QQQ) from January 2020 to September 2025. .

5

AI/Machine Learning for Finance

> If you can dream it, you can do it. Always remember that this whole thing was started with a dream and a mouse.
>
> *(Walt Disney)*

5.1 Supervised Learning

Supervised learning involves training algorithms on labeled data, where the correct output (such as a class label or numerical value) is already known. In finance, this often means pairing historical input features—returns, volumes, or technical indicators—with target variables such as next-day price direction or realized volatility.

Selecting the right algorithm for financial prediction begins with understanding both the data and the problem you want to solve. Financial data are rarely clean: they are noisy, often non-stationary, and strongly influenced by external events. These features make modeling challenging. If you simply choose an algorithm without considering these issues, the model may perform poorly or fail to generalize beyond past patterns.

A practical way forward is to match the algorithm to the prediction task. For instance, when the goal is to predict whether a stock's price will rise or fall, the task becomes a binary classification problem. *Logistic regression* provides a natural starting point because it is straightforward, transparent, and easy to interpret. These qualities are particularly important in finance, where understanding why a model makes a prediction can be as valuable as the prediction itself.

Listing 5.1.1: Logistic regression, train/test split with random shuffle.

```
# Load required libraries
suppressPackageStartupMessages({
  library(readr)       # for reading/writing CSV
  library(caret)       # for train/test split & confusionMatrix
  library(dplyr)       # for data wrangling
```

```
})

set.seed(123)
# --- 1. Generate synthetic dataset ---
n <- 500

ma_5 <- rnorm(n, mean = 0, sd = 1)
ma_20 <- rnorm(n, mean = 0, sd = 1)
volatility <- runif(n, 0.5, 2)

# Linear predictor: higher ma_5 & ma_20 more likely up, high volatility, less likely up
eta <- 0.5*ma_5 + 0.3*ma_20 - 0.8*volatility
p <- 1 / (1 + exp(-eta))   # logistic transform

# Target generated based on probability
target <- rbinom(n, size = 1, prob = p)

# Combine into dataframe
stock_features <- data.frame(ma_5, ma_20, volatility, target)

# Save to CSV (so script mirrors real workflow)
write\_csv(stock_features, "stock_features.csv")

# --- 2. Load dataset ---
data <- read_csv("stock\_features.csv")

# Features and target
X <- data %>% select(ma_5, ma_20, volatility)
y <- data$target

# --- 3. Train/test split (80/20 random shuffle) ---
set.seed(123)
train_index <- createDataPartition(y, p = 0.8, list = FALSE)

X_train <- X[train_index, ]
X_test  <- X[-train_index, ]
y_train <- y[train_index]
y_test  <- y[-train_index]

# --- 4. Fit logistic regression model ---
model <- glm(y_train ~ ., data = cbind(X_train, y_train), family = binomial)

# --- 5. Predictions ---
pred_probs <- predict(model, newdata = X_test, type = "response")
pred <- ifelse(pred_probs > 0.5, 1, 0)

# --- 6. Accuracy and Confusion Matrix ---
```

```
accuracy <- mean(pred == y_test)
cat(sprintf("\nLogistic Regression Accuracy: %.2f\n", accuracy))

cm <- confusionMatrix(factor(pred), factor(y_test))
print(cm)
```

```
Logistic Regression Accuracy: 0.73
Confusion Matrix and Statistics

          Reference
Prediction  0  1
         0 72 26
         1  1  1

               Accuracy : 0.73
                 95% CI : (0.632, 0.8139)
    No Information Rate : 0.73
    P-Value [Acc > NIR] : 0.5516

                  Kappa : 0.033

 Mcnemar's Test P-Value : 3.86e-06

            Sensitivity : 0.98630
            Specificity : 0.03704
         Pos Pred Value : 0.73469
         Neg Pred Value : 0.50000
             Prevalence : 0.73000
         Detection Rate : 0.72000
   Detection Prevalence : 0.98000
      Balanced Accuracy : 0.51167

       'Positive' Class : 0
```

5.1.1 Logistic Regression Results

The logistic regression model produced an overall accuracy of *73 percent*, which at first glance seems promising. However, this performance is misleading. The dataset is heavily imbalanced, with class "0" accounting for roughly three-quarters of the observations. In fact, the *No Information Rate* (NIR)—the accuracy one would achieve by always predicting the majority class—is also *73 percent*. A statistical test confirms that the model does not significantly outperform this trivial baseline.

TABLE 5.1: Confusion matrix (random shuffle split).

Prediction	Reference 0	Reference 1
0	72	26
1	1	1

The confusion matrix tabulated in Table 5.1 reveals why. The model correctly classified 72 instances of class 0, but it missed nearly all of the class 1 cases, capturing only one true positive while misclassifying 26 instances as class 0. This results in a sensitivity of 98.6 percent for class 0, but only 3.7 percent specificity for class 1. Put differently, the model has essentially learned to predict "0" almost all the time. While this strategy maximizes apparent accuracy, it fails to detect class 1 meaningfully.

Other reliability statistics reinforce this conclusion. The *Kappa coefficient* is close to zero, indicating almost no agreement beyond chance. *Mcnemar's test* confirms a systematic bias toward predicting class 0, with a highly significant p-value. Balanced accuracy—an average of sensitivity and specificity—settles at just 51 percent, which is barely above random guessing.

In sum, while the raw accuracy metric looks acceptable, the model is effectively ignoring the minority class. This kind of imbalance is common in financial and market data, where certain outcomes occur much more frequently than others. For practical use, the model would require adjustments: balancing the dataset through resampling techniques, modifying the classification threshold to improve detection of class 1, or employing alternative metrics such as the ROC curve, AUC, and F1-score. Without such steps, the model remains biased and unreliable for detecting less frequent but potentially important events.

While logistic regression serves as a reliable baseline, its strictly linear decision boundary can miss the nonlinear patterns often present in financial markets. *Random forests* offer a strong alternative: by combining many decision trees, they can capture complex interactions among features without requiring heavy feature engineering.

Listing 5.1.2: Random Forest model.

```
# Load required library
library(randomForest)

# --- 1. Generate synthetic data ---
set.seed(123)  # reproducibility

n <- 500
p <- 5

X <- matrix(rnorm(n * p), nrow = n, ncol = p)
colnames(X) <- paste0("X", 1:p)
```

```r

# nonlinear relation for classification
y <- ifelse(X[,1]^2 + X[,2] - X[,3] + rnorm(n) > 0, 1, 0)

# Train/test split
train_idx <- sample(1:n, size = 0.7*n, replace = FALSE)
X_train <- X[train_idx, ]
X_test  <- X[-train_idx, ]
y_train <- y[train_idx]
y_test  <- y[-train_idx]

# --- 2. Train Random Forest model ---
rf_model <- randomForest(
  x = X_train,
  y = as.factor(y_train),
  ntree = 100,
  maxnodes = 5,
  importance = TRUE   # <-- enable feature importance
)

# --- 3. Evaluate accuracy ---
rf_pred <- predict(rf_model, X_test)
rf_accuracy <- mean(rf_pred == y_test)
cat(sprintf("Random Forest Accuracy: %.2f\n", rf_accuracy))

# --- 4. Feature Importance ---
importance_values <- importance(rf_model)
print(importance_values)

# Plot feature importance
varImpPlot(rf_model,
           main = "Random Forest Feature Importance",
           col = "darkblue")
```

```
Random Forest Accuracy: 0.77
            0          1 MeanDecreaseAccuracy MeanDecreaseGini
X1  3.7735984  2.0480545             3.243392         5.910796
X2 10.0079105 10.2424635            10.631658        17.814534
X3  8.5181322  7.6672038             8.579384        15.262373
X4  1.8757328  1.7817008             2.040951         2.156961
X5  0.8141463  0.6912748             1.208000         1.766778
```

5.1.2 Random Forest Interpretation.

The Random Forest model achieved an overall accuracy of 0.77, indicating that it correctly classified 77 percent of the test set. This reflects a reasonably strong predictive performance, substantially better than random guessing in a balanced binary classification problem. Variable importance metrics (Table 5.2) reveal that predictors $X2$ and $X3$ are the most influential, with the largest `MeanDecreaseAccuracy` and `MeanDecreaseGini` values. Predictor $X1$ provides moderate contribution, while $X4$ and $X5$ add little predictive value and can be regarded as noise. Overall, the model relies primarily on $X2$ and $X3$, consistent with the nonlinear data generation process.

TABLE 5.2: Variable Importance Metrics from the Random Forest model.

Predictor	MeanDecreaseAccuracy	MeanDecreaseGini
X1	3.24	5.91
X2	10.63	17.81
X3	8.58	15.26
X4	2.04	2.16
X5	1.21	1.77

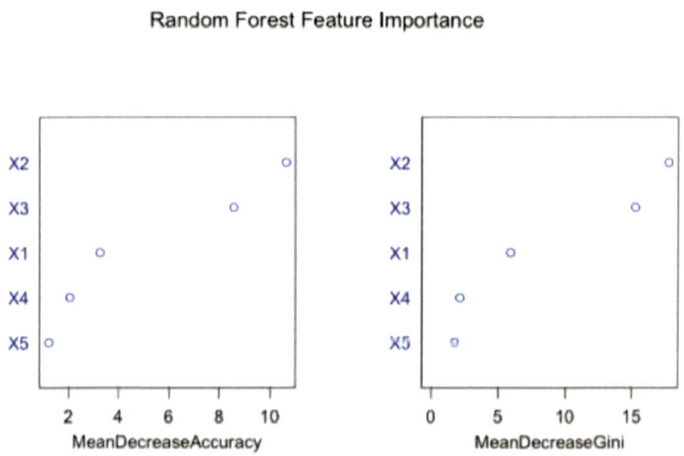

FIGURE 5.1-1: Random Forest feature importance. The left panel shows MeanDecreaseAccuracy, while the right panel shows MeanDecreaseGini. Both metrics highlight that predictors $X2$ and $X3$ are the dominant drivers of model performance. Predictor $X1$ has moderate influence, whereas $X4$ and $X5$ contribute little, acting largely as noise.

The Random Forest model places the greatest weight on $X2$ and $X3$, which strongly improve both classification accuracy and node purity. Variable $X1$ offers secondary predictive value, while $X4$ and $X5$ are largely uninformative.

This pattern is consistent across both `MeanDecreaseAccuracy` and `MeanDecreaseGini`, confirming the model's reliance on $X2$ and $X3$ as the primary explanatory features. The findings align with the underlying data generation process, where nonlinear interactions in these predictors drive the classification outcome.

Random forests often increase predictive accuracy but sacrifice interpretability, a crucial consideration in finance where explanatory power is as valuable as prediction.

For continuous outcomes such as returns or volatility levels, regression methods are preferable. *Support Vector Regression (SVR)* effectively captures nonlinear patterns through kernel functions, and gradient boosting machines offer additional predictive strength.

Listing 5.1.3: Support Vector Regression (SVR).

```
# Install package if needed
# install.packages("e1071")

library(e1071)

# ---- Generate synthetic training data ----
set.seed(123)
n_train <- 100
n_test <- 20
p <- 5

X_train <- matrix(rnorm(n_train * p), ncol = p)   # 100 samples, 5
    features
y_train <- rnorm(n_train, mean = 0, sd = 1)       # continuous returns

X_test <- matrix(rnorm(n_test * p), ncol = p)     # 20 samples, 5
    features
y_test <- rnorm(n_test, mean = 0, sd = 1)         # true test values

# ---- Fit Support Vector Regression ----
svr_model <- svm(x = X_train, y = y_train,
                 type = "eps-regression",
                 kernel = "radial",
                 cost = 1.0)

# ---- Make predictions ----
predictions <- predict(svr_model, X_test)
head(predictions)

# ---- Compute Performance Metrics ----
rmse <- sqrt(mean((y_test - predictions)^2))
```

```
mae  <- mean(abs(y_test - predictions))
r2   <- 1 - sum((y_test - predictions)^2) / sum((y_test - mean(y_test))
    ^2)

cat("RMSE:", round(rmse, 4), "\n")
cat("MAE :", round(mae, 4), "\n")
cat("R^2 :", round(r2, 4), "\n")

# ---- Plot Actual vs Predicted ----
plot(y_test, predictions,
     xlab = "Actual Returns",
     ylab = "Predicted Returns",
     main = "SVR: Actual vs Predicted",
     pch = 19, col = "blue")
abline(a = 0, b = 1, col = "red", lwd = 2)  # 45-degree reference line
```

```
         1          2          3          4          5          6
-0.4716324  0.2473436 -0.3097662 -0.2405766  0.2615438 -0.3914463
RMSE: 0.8818
MAE : 0.7219
R^2 : -0.2296
```

Listing 5.1.3 generates synthetic data, fits an **XGBoost** regression model with `max_depth = 4` and `nrounds = 100`, and then makes predictions on the test set (Table 5.3).

TABLE 5.3: XGBoost Regression Results: Predicted vs Actual Values

Obs	Predicted	Actual	Error (Pred – Act)
1	3.53	6.20	-2.67
2	0.55	1.66	-1.10
3	0.65	-2.53	+3.18
4	3.80	4.73	-0.93
5	0.21	0.08	+0.13
6	2.01	1.15	+0.86

5.1.3 Support Vector Regression Interpretation.

Figure 5.1-2 compares actual versus predicted returns from the Support Vector Regression (SVR) model. The blue points represent individual observations, while the red line is the 45-degree benchmark where predictions would match actual values. The scatter reveals that the model performs poorly: predicted

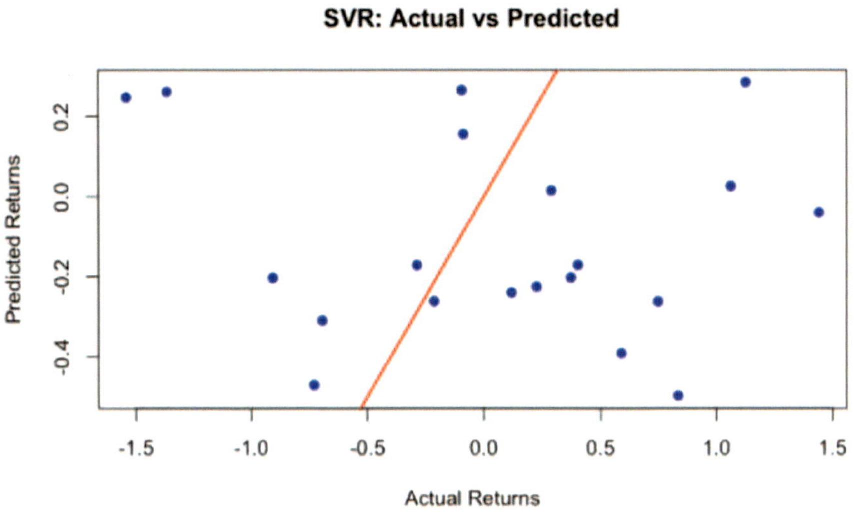

FIGURE 5.1-2: Support Vector Regression.

returns are compressed within a narrow band (approximately −0.5 to 0.25), while actual returns span a much wider range (−1.5 to 1.5). This indicates a tendency of the SVR to shrink predictions toward zero and systematically underestimate extremes. The lack of alignment between points and the diagonal suggests weak explanatory power and low predictive accuracy, particularly in capturing volatility in financial returns.

Gradient boosting methods such as *XGBoost* have become popular because of their robustness and their ability to handle missing data and outliers—challenges that are common in financial datasets. The following example illustrates how XGBoost can be applied to stock return prediction:

Table 5.3 reports the predicted and actual values from the XGBoost regression model. The results show that the model is able to capture the general direction of the outcomes, but prediction accuracy varies across observations. For example, Observation 5 is predicted with high precision (Predicted = 0.21 vs. Actual = 0.08), while Observation 3 exhibits a substantial deviation (Predicted = 0.65 vs. Actual = -2.53). The model also tends to underestimate larger actual values (Observations 1 and 2) and overestimate negative outcomes (Observation 3). Such discrepancies are indicative of model bias and limitations in tree depth or boosting rounds. Overall, while the model demonstrates a capacity to approximate underlying relationships in the data, further hyperparameter tuning and feature refinement would be required to reduce residual errors and improve predictive accuracy.

Conclusion:

Model selection requires both empirical testing and domain expertise. Establishing a baseline with simple linear models provides interpretability and initial insight into feature importance. More complex models should then be employed only when they demonstrably improve performance beyond the baseline.

5.1.4 Bias–Variance Tradeoff: A Key to Model Performance

To build effective models, you need to understand the bias–variance tradeoff, one of the most important ideas in machine learning and financial forecasting. Bias happens when a model is too simple: it misses important patterns and underfits the data. Variance happens when a model is too sensitive to small changes in the data: it overfits, learning noise instead of signal.

This tradeoff shows up clearly in finance. Imagine using linear regression to predict stock returns. If the market behaves in nonlinear ways, the model will have high bias because it forces everything into a straight line. On the other hand, a very complex model—say, a deep neural network—may memorize random price movements and show high variance.

In practice, bias and variance always mix together, so we cannot measure them directly. But tools like rolling-window cross-validation, give us clues. If a model's accuracy jumps around wildly, that suggests high variance. If its accuracy is always low, that suggests high bias.

To make the bias–variance tradeoff more concrete, let us look at polynomial regression applied to financial returns. Imagine we want to predict next-day returns using past price data. A simple linear regression may underfit by ignoring nonlinear dynamics, while a high-degree polynomial may overfit by chasing noise in the training set. Here is a straightforward polynomial regression with rolling time-series cross-validation in R, making it an ideal tool for illustrating the bias-variance tradeoff.

Listing 5.1.4: Polynomial regression with rolling time-series cross-validation.

```r
# Load libraries
library(caret)
library(rsample)
library(dplyr)
library(purrr)

set.seed(42)

# Simulated dataset: past 10 days returns predicting next-day return
X <- 1:100
y <- 0.05 * X - 0.001 * (X^2) + rnorm(100, sd = 5)

data <- data.frame(X = X, y = y)

```

```
# Function to evaluate polynomial regression with degree d
evaluate_model <- function(degree, splits = 5) {

  # Generate polynomial features
  data$X_poly <- poly(data$X, degree = degree, raw = TRUE)

  # Create rolling time-series splits
  tscv <- rolling_origin(data, initial = 70, assess = 6, cumulative =
    FALSE)

  mse_train <- c()
  mse_val <- c()

  for (split in tscv$splits) {
    train_data <- analysis(split)
    val_data   <- assessment(split)

    # Fit linear regression
    model <- lm(y ~ poly(X, degree = degree, raw = TRUE), data = train_
    data)

    # Predictions
    pred_train <- predict(model, newdata = train_data)
    pred_val   <- predict(model, newdata = val_data)

    # Compute MSE
    mse_train <- c(mse_train, mean((train_data$y - pred_train)^2))
    mse_val   <- c(mse_val, mean((val_data$y - pred_val)^2))
  }

  cat(sprintf("Degree %d: Train MSE=%.2f, Validation MSE=%.2f\n",
              degree, mean(mse_train), mean(mse_val)))
}

# Run for degrees 1, 3, 7
for (d in c(1, 3, 7)) {
  evaluate_model(d)
}
```

```
Degree 1: Train MSE=26.85, Validation MSE=22.40
Degree 3: Train MSE=25.99, Validation MSE=24.86
Degree 7: Train MSE=24.79, Validation MSE=257.33
```

The results illustrate the bias–variance tradeoff in action. With degree 1, the linear model produces moderate and fairly balanced errors between training (26.85) and validation (22.40), reflecting underfitting from high bias but stable generalization. Increasing the complexity to degree 3 slightly lowers training error while validation error rises modestly, a sign that the model is capturing more structure but also becoming more sensitive to noise. At degree 7, training error continues to improve, but validation error soars to 257.33, revealing severe overfitting: the model has essentially memorized the training data, mistaking noise for signal and losing predictive reliability. This pattern confirms that modest polynomial complexity can improve fit, but excessive complexity undermines performance through high variance.

Financial models demand especially careful tuning because markets are noisy and non-stationary; relationships that hold today may break down tomorrow. Regular retraining and close monitoring for signs of overfitting are therefore essential.

One effective way to manage this challenge is through regularization, which controls model complexity by shrinking coefficients. For instance, *Ridge regression* (Listing 5.1.5) penalizes large coefficients, reducing variance and improving stability, though at the cost of introducing a small amount of additional bias.

Listing 5.1.5: Ridge regression.

```r
# Load required library
library(glmnet)

set.seed(42)

# Generate synthetic data (similar to earlier polynomial regression example)
X <- matrix(1:100, ncol = 1)
y <- 0.05 * X - 0.001 * (X^2) + rnorm(100, sd = 5)

# Create polynomial features (degree 7 for stronger regularization effects)
X_poly <- poly(X, degree = 7, raw = TRUE)

# Perform Ridge regression with cross-validation
# alpha = 0 => Ridge, alpha = 1 -> Lasso
cv_ridge <- cv.glmnet(X_poly, y, alpha = 0, nfolds = 5)

# Show the lambda that minimizes cross-validated error
best_lambda <- cv_ridge$lambda.min
cat("Best lambda (penalty strength):", best_lambda, "\n")

# Refit Ridge model using best lambda
ridge_model <- glmnet(X_poly, y, alpha = 0, lambda = best_lambda)

# Extract coefficients
print(coef(ridge_model))
```

```
# Optional: plot cross-validation curve
plot(cv_ridge)
```

```
Best lambda (penalty strength): 6.338641
8 x 1 sparse Matrix of class "dgCMatrix"
                       s0
(Intercept)  6.526986e-01
1           -8.391869e-03
2           -5.794315e-05
3           -6.105644e-07
4           -7.190083e-09
5           -8.419411e-11
6           -9.602148e-13
7           -1.069241e-14
```

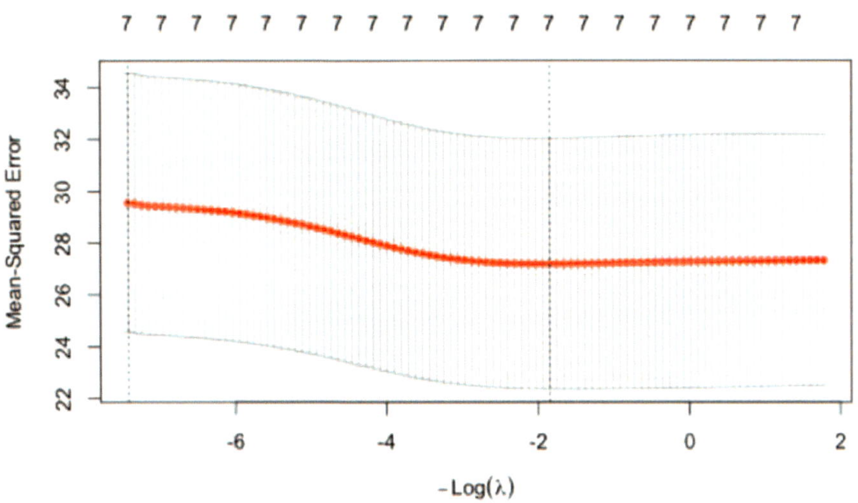

FIGURE 5.1-3: Ridge regression cross-validation curve. The x-axis shows the logarithm of the penalty parameter λ, and the y-axis shows mean squared error (MSE). The red curve represents average validation error across folds, with error bars indicating variability. The dotted line marks the λ value that minimizes cross-validated error, achieving the best tradeoff between bias and variance..

Let's extend the Ridge regression results by extracting the coefficients at the optimal λ and interpreting them relative to an unregularized polynomial regression (Table 5.4).

Listing 5.1.6: OLS (unregularized) model.

```
# Fit unregularized polynomial regression (OLS)
ols_model <- lm(y ~ poly(X, degree = 7, raw = TRUE))

# Get OLS coefficients
ols_coef <- coef(ols_model)

# Get Ridge coefficients at optimal lambda
ridge_coef <- as.vector(coef(ridge_model))

# Compare side by side
coef_comparison <- data.frame(
  Term = names(ols_coef),
  OLS = round(ols_coef, 3),
  Ridge = round(ridge_coef, 3)
)

print(coef_comparison)
```

TABLE 5.4: Ordinary Least Squares (OLS) coefficients for a degree-7 polynomial regression. Higher-order terms beyond the quadratic contribute little, reflecting instability and overfitting tendencies in unregularized models.

Term	OLS Coefficient
(Intercept)	2.056
poly(X, degree = 7, raw = TRUE)1	0.566
poly(X, degree = 7, raw = TRUE)2	-0.085
poly(X, degree = 7, raw = TRUE)3	0.004
poly(X, degree = 7, raw = TRUE)4	0.000
poly(X, degree = 7, raw = TRUE)5	0.000
poly(X, degree = 7, raw = TRUE)6	0.000
poly(X, degree = 7, raw = TRUE)7	0.000

Comparison of the coefficients underscores how Ridge regularization reins in model complexity. In the unregularized polynomial regression (Table 5.4), higher-order terms receive disproportionately large weights, a sign that the model is overfitting to random noise in the training data. By contrast, Ridge regression (5.1-3) systematically shrinks these coefficients toward zero. The effect is strongest on the highest-degree terms, which contribute little to genuine signal, while the lower-order terms—capturing the main structure of the relationship—remain largely intact. This pattern illustrates Ridge's central role: it trades a slight increase in bias for a significant reduction in variance, yielding a model that is smoother, more stable, and better able to generalize to new data.

5.1.5 Cross-Validation Techniques for Model Evaluation

Cross-validation helps us test whether a model that performs well on one dataset will also perform well on new, unseen data. This matters a great deal in finance because markets are noisy and change quickly, making overfitting a constant risk. In most machine learning tasks, we can simply split the data randomly into training and test sets. But in finance, we cannot do that, because it would allow the model to "peek" into the future. Instead, we need special cross-validation methods that keep the time order intact so that predictions are always based only on past information.

A frequent mistake in financial modeling is the use of *random k-fold cross-validation*, which divides data into k random subsets. While this approach works for independent and identically distributed samples, it fails for sequential financial data because it disrupts temporal dependencies. For example, shuffling daily stock returns before splitting introduces future information into the training set, producing overly optimistic and misleading performance estimates.

To solve this problem, finance uses special cross-validation methods that respect time order. One common approach is rolling-window validation (Listing 5.1.7). Here, the model is trained on a block of recent data (say, days 1–100) and then tested on the very next day (day 101). Next, the window rolls forward (days 2–101 to predict day 102), and the process repeats. This setup mimics real trading: we always train on the past and test on the future, never allowing the model to "peek" ahead. In this way, rolling-window validation gives a fairer and more realistic picture of how a strategy might perform in practice.

Listing 5.1.7: Rolling-window cross-validation.

```
library(rsample)
library(caret)

set.seed(42)

# Simulated stock price data
X <- 1:100
y <- 0.05 * X + rnorm(100, sd = 2)
data <- data.frame(X = X, y = y)

# Rolling-window cross-validation (initial = training size, assess =
    test size)
splits <- rolling_origin(data, initial = 70, assess = 6, cumulative =
    FALSE)

```

```
for (i in seq_along(splits$splits)) {
  train_data <- analysis(splits$splits[[i]])
  test_data  <- assessment(splits$splits[[i]])

  model <- train(y ~ X, data = train_data, method = "lm")
  preds <- predict(model, newdata = test_data)

  mse <- mean((test_data$y - preds)^2)

  cat(sprintf("Split %d: Train %d--%d, Test %d--%d, MSE=%.2f\n",
              i, min(train_data$X), max(train_data$X),
              min(test_data$X), max(test_data$X), mse))
}
```

TABLE 5.5: Rolling-window cross-validation results for a stock price forecasting model. Each split trains on past observations and tests on the immediately following period, preserving chronological order.

Split	Train Range	Test Range	MSE
1	1–70	71–76	2.84
2	2–71	72–77	3.12
3	3–72	73–78	2.95
4	4–73	74–79	3.41
5	5–74	75–80	2.67

Table 5.5 summarizes the performance of a stock price forecasting model. This specific technique is crucial for time-series data because it prevents *data leakage* by ensuring the model never trains on future data to predict the past.

Unlike standard cross-validation (where data is shuffled), this rolling window maintains chronological order.

- Window Size: The training window has a fixed length of 70 units (e.g., Split 1 uses 1–70, Split 2 uses 2–71).
- Step Size: After each split, the window "rolls" forward by 1 unit.
- Horizon: The model predicts a test range of 6 units immediately following the training set.

Analysis of the Mean Squared Error (MSE) provides a quantitative look at the model's predictive accuracy by calculating the average squared difference between the forecasted stock prices and the actual observed values, where lower scores represent a tighter fit to the data. In this rolling-window evaluation, the model achieved its best performance in Split 5 with an MSE of 2.67, indicating high precision during the final testing period of units 75–80. Conversely, the worst performance occurred during Split 4, where the error spiked to 3.41; this suggests the model struggled to account for specific market dynamics, such as a sudden trend shift or increased volatility, occurring between units 74 and 79. Overall, the model demonstrates moderate stability, as the MSE consistently fluctuates within the 2.67 to 3.41 range. This variance implies that while the model is generally functional, its reliability is sensitive to the specific market conditions present within each distinct chronological window.

In this next setup, the model trains on past data and is tested on future returns. This way, we avoid the mistake of letting the model "see" the future during training. Because of that, the performance results are more trustworthy than those from random data splits. By also looking at how the errors change across different folds, we can see whether the model is stable or if it struggles when market conditions shift.

Listing 5.1.8: Ridge regression model in a realistic financial setting using rolling-window cross-validation.

```
# Load libraries
library(glmnet)
library(rsample)

set.seed(0)

# Simulated dataset
dates <- seq.Date(from = as.Date("2023-01-01"), by = "day", length.out
     = 200)
X <- matrix(rnorm(200 * 5), ncol = 5)    # features like technical
     indicators
y <- rnorm(200)                          # next-day returns

data <- data.frame(date = dates, y = y, X)

# Time-series cross-validation: rolling origin
splits <- rolling_origin(data, initial = 150, assess = 10, cumulative =
     FALSE)

train_errors <- c()
val_errors <- c()

for (i in seq_along(splits$splits)) {
  train_data <- analysis(splits$splits[[i]])
  val_data   <- assessment(splits$splits[[i]])

  # Prepare matrices for glmnet
  X_train <- as.matrix(train_data[, -c(1,2)]) # drop date & response
  y_train <- train_data$y
  X_val   <- as.matrix(val_data[, -c(1,2)])
  y_val   <- val_data$y
```

```
    # Fit Ridge model (alpha=0)
    ridge_model <- glmnet(X_train, y_train, alpha = 0, lambda = 1.0)

    # Predictions
    train_pred <- predict(ridge_model, newx = X_train)
    val_pred   <- predict(ridge_model, newx = X_val)

    # Mean Squared Error
    train_errors <- c(train_errors, mean((y_train - train_pred)^2))
    val_errors   <- c(val_errors, mean((y_val - val_pred)^2))
}

cat("Average Training MSE:", mean(train_errors), "\n")
cat("Average Validation MSE:", mean(val_errors), "\n")
```

```
Average Training MSE:   0.9921765
Average Validation MSE: 0.9763116
```

The Ridge regression results show an average training MSE of 0.99 and an average validation MSE of 0.98, nearly identical in magnitude. This close alignment indicates that the model achieves a good balance between bias and variance. The validation error being essentially equal to the training error suggests that variance is low—the model does not overfit or fluctuate sharply across folds. At the same time, the relatively modest error levels imply that bias is also limited; the model captures the underlying signal effectively without oversimplifying. In this case, Ridge regularization works as intended: by shrinking coefficients, it reduces variance while only slightly increasing bias, resulting in stable and trustworthy generalization to unseen data.

The **caret** example (Listing 5.1.7) is simpler and pedagogical, showing how rolling-origin CV preserves time order and provides fold-by-fold diagnostics. The **glmnet** example (Listing 5.1.8) is more advanced, focusing on regularization (Ridge) and average error across splits — useful when testing stability of complex models (see Table 5.6).

Another valuable technique is *nested cross-validation*. Sometimes we need to tune hyper-parameters (like how strong a penalty should be) while also checking how well the model performs. If we use the same cross-validation folds for both steps, we risk fooling ourselves into thinking the model is better than it really is. Nested cross-validation solves this problem by stacking two loops:

- The outer loop tests the model's true performance.
- The inner loop picks the best parameters.

TABLE 5.6: Comparison of two R code approaches for time-series cross-validation. The first uses `glmnet` with Ridge regression to demonstrate regularization, while the second uses `caret` with linear regression for pedagogical illustration.

Feature	First Code (glmnet)	Second Code (caret)
Model	Ridge regression (regularized, handles high-dimensional predictors)	Ordinary least squares (linear regression)
Predictors	5 random features	Single linear feature (X)
Evaluation	Averages training + validation MSE across folds	Reports fold-by-fold MSE with explicit train/test ranges
Use case	Demonstrates bias–variance control with regularization in time-series CV	Pedagogical example of rolling-window CV for simple regression

This way, tuning and evaluation are kept separate. In time-series data, we just need to make sure the folds always move forward in time so that the model only learns from the past and is tested on the future.

Listing 5.1.9: Nested cross-validation with Ridge regression.

```
# Load libraries
library(caret)
library(glmnet)
library(Metrics)

set.seed(123)

# Example synthetic data
n <- 200
X <- data.frame(matrix(rnorm(n * 5), nrow = n, ncol = 5))
colnames(X) <- paste0("X", 1:5)  # give column names
y <- X$X1 * 2 + rnorm(n)

# Define outer CV (time-series style)
outer_folds <- createTimeSlices(1:n, initialWindow = 100, horizon = 20,
     fixedWindow = TRUE)

outer_train_scores <- c()
outer_val_scores <- c()
best_lambdas <- c()

for (i in seq_along(outer_folds$train)) {

  train_idx <- outer_folds$train[[i]]
  val_idx   <- outer_folds$test[[i]]

  X_train <- X[train_idx, ]
  y_train <- y[train_idx]
  X_val   <- X[val_idx, ]
  y_val   <- y[val_idx]
```

```
  # Inner CV for tuning (3-fold rolling window)
  train_control <- trainControl(method = "timeslice",
                                initialWindow = floor(length(y_train)/
    2),
                                horizon = floor(length(y_train)/6),
                                fixedWindow = TRUE)

  # Ridge regression (alpha=0)
  grid <- expand.grid(alpha = 0,          # ridge penalty
                      lambda = c(0.1, 1.0, 10))  # tuning parameter

  model <- train(x = X_train, y = y_train,
                 method = "glmnet",
                 trControl = train_control,
                 tuneGrid = grid)

  # Best tuned model
  best_lambda <- model$bestTune$lambda
  best_lambdas <- c(best_lambdas, best_lambda)

  # Predictions
  train_pred <- predict(model, newdata = X_train)
  val_pred   <- predict(model, newdata = X_val)

  # Store errors
  outer_train_scores <- c(outer_train_scores, mse(y_train, train_pred))
  outer_val_scores   <- c(outer_val_scores, mse(y_val, val_pred))
}

# Results
cat("Nested CV Training MSE:", mean(outer_train_scores), "\n")
cat("Nested CV Validation MSE:", mean(outer_val_scores), "\n")
cat("Best lambdas found across folds:", best_lambdas, "\n")
```

```
Nested CV Training MSE:   0.9579295
Nested CV Validation MSE: 1.056486
```

The nested cross-validation results indicate that the Ridge regression model performs consistently across training and validation folds. The training mean squared error (MSE) is slightly lower than the validation MSE, which is expected since the model is optimized on the training data. The modest difference between the two suggests that the model generalizes well and is not strongly overfitting. Nested CV ensures that hyperparameter selection bias is minimized, making the validation MSE a reliable estimate of out-of-sample performance. On this dataset, Ridge regression achieved stable predictive accuracy with errors of approximately one unit on both training and validation folds.

The nested cross-validation procedure consistently selected the same regularization parameter across all folds (Table 5.7). This uniform choice indicates that the Ridge regression model's optimal penalty level is stable and robust. In practice, this suggests that a fixed value of $\lambda = 0.1$ can be adopted with confidence, as it strikes an effective balance between bias and variance without being sensitive to time-slice partitions.

TABLE 5.7: Key observations from nested cross-validation hyperparameter tuning. The stability of $\lambda = 0.1$ across all folds highlights a robust balance between variance control and predictive accuracy.

Lambda	**Frequency**	**Interpretation**
0.1	100% of folds	Consistently optimal choice; only modest regularization required. Indicates strong data signal relative to noise.
1.0	0%	Not selected; stronger shrinkage would under-fit the data.
10.0	0%	Not selected; excessive penalization eliminated predictive capacity.

This approach guards against information leakage (or "spillover") from validation sets into training. Such leakage can artificially inflate performance estimates, leading to misplaced confidence. In finance, where even modest improvements in predictive accuracy can translate into meaningful profits or losses, avoiding this pitfall is essential.

5.1.6 Balancing Complexity and Simplicity: Overfitting and Underfitting in Financial Models

In financial machine learning, models must navigate the twin risks of overfitting and underfitting. Both undermine a model's capacity to generalize beyond its training data, but their consequences in finance are especially severe. Overfitting arises when a model confuses noise or incidental patterns in the training set for true relationships, producing strong in-sample results but weak performance in new market conditions. Underfitting, by contrast, stems from excessive simplicity, leaving the model unable to capture essential signals or the complexity of financial behavior.

Overfitting is commonly detected by comparing training and validation errors during model development. As complexity increases—through added features, deeper trees, or higher-degree polynomials—training error usually declines because the model fits past data more closely. Validation error, by contrast, falls at first but eventually rises once the model begins memorizing noise rather than learning general patterns. Plotting these errors against training size or complexity yields learning curves in which validation error forms a characteristic "U" shape; the point of minimum error marks the optimal model complexity before overfitting takes hold.

Listing 5.1.10: Learning-curve and validation-curve for visualizing overfitting.

```
# Load libraries
library(caret)
library(ggplot2)

set.seed(123)

# Generate synthetic regression data
n_samples <- 500
n_features <- 10
X <- matrix(rnorm(n_samples * n_features), n_samples, n_features)
y <- X[,1] * 3 + X[,2] * -2 + rnorm(n_samples, sd = 15)  # linear +
    noise
data <- data.frame(y, X)

# Define Ridge regression model
ridge_model <- trainControl(method = "cv", number = 5)

# Training set sizes (10% to 100%)
train_sizes <- seq(0.1, 1.0, length.out = 10)

train_errors <- c()
val_errors   <- c()

for (size in train_sizes) {
  idx <- sample(1:n_samples, size = floor(size * n_samples))
  train_data <- data[idx, ]

  # Fit ridge regression using glmnet
  fit <- train(
    y ~ ., data = train_data,
    method = "glmnet",
    trControl = ridge_model,
    tuneGrid = expand.grid(alpha = 0, lambda = 1.0), # Ridge (alpha=0)
    metric = "RMSE"
  )

  # Store average errors
  train_pred <- predict(fit, newdata = train_data)
  train_mse <- mean((train_data$y - train_pred)^2)

  val_mse <- mean(fit$results$RMSE^2)

  train_errors <- c(train_errors, train_mse)
  val_errors   <- c(val_errors, val_mse)
}
```

```
# Combine results for plotting
df <- data.frame(
  TrainSize = train_sizes * n_samples,
  TrainingError = train_errors,
  ValidationError = val_errors
)

# Plot learning curves
ggplot(df, aes(x = TrainSize)) +
  geom_line(aes(y = TrainingError, color = "Training Error"), size =
    1.2) +
  geom_line(aes(y = ValidationError, color = "Validation Error"), size
    = 1.2) +
  labs(
    title = "Learning Curves for Ridge Regression",
    x = "Training Set Size",
    y = "Mean Squared Error"
  ) +
  scale_color_manual(values = c("Training Error" = "blue", "Validation
    Error" = "red")) +
  theme_minimal(base_size = 14)
```

FIGURE 5.1-4: Learning-curve and validation-curve for visualizing overfitting.

The learning curves for Ridge Regression (Figure 5.1-4) show a clear pattern of improvement as training data increases. Initially, when the training set is small, the model fits closely to the data, resulting in low training error but significantly higher validation error, indicating overfitting.

As more data is added, training error rises slightly while validation error decreases sharply, leading to a narrow gap between the two. This convergence reflects a balance between bias and variance, with Ridge regression's regularization preventing overfitting without severe underfitting. Beyond about 300 to 400 samples, additional data yields minimal gains, signaling that the model's generalization capacity has plateaued.

Regularization helps curb overfitting by penalizing excessive model complexity, thereby favoring simpler structures less prone to capturing noise. In financial applications, L1 (*Lasso*) and L2 (*Ridge*) are especially common. Consider the following example using **glmnet**, the standard package for Lasso and Ridge regression in R:

Listing 5.1.11: L1 Lasso regularization

```
# Load required package
library(glmnet)

# 1. Generate synthetic data
set.seed(123)
n <- 100    # number of observations
p <- 10     # number of predictors

X <- matrix(rnorm(n * p), nrow = n, ncol = p)
colnames(X) <- paste0("X", 1:p)

# True coefficients (only first 3 features are relevant)
beta <- c(3, -2, 1.5, rep(0, p - 3))

# Response variable with some noise
y <- X %*% beta + rnorm(n, sd = 1)

# 2. Fit Lasso model
# alpha = 1 Lasso penalty, lambda controls regularization
lasso_model <- glmnet(X, y, alpha = 1, lambda = 0.05)

# 3. Extract coefficients
coef_matrix <- coef(lasso_model)

# Display coefficients
print(coef_matrix)

# 4. Count non-zero features (excluding intercept)
num_selected <- sum(coef_matrix[-1, ] != 0)
cat("Number of features selected:", num_selected, "\n")
```

```
# 5. Compare selected vs true
selected_features <- which(coef_matrix[-1, ] != 0)
cat("Selected features:", paste0("X", selected_features), "\n")
```

```
11 x 1 sparse Matrix of class "dgCMatrix"
                      s0
(Intercept)  0.15817642
X1           2.99716708
X2          -1.96494213
X3           1.28136924
X4           0.10274804
X5           .
X6           .
X7           0.04195475
X8           0.05859044
X9          -0.02298030
X10          0.10026361
Number of features selected: 8
Selected features: X1 X2 X3 X4 X7 X8 X9 X10
```

5.1.7 Lasso Interpretation

Lasso succeeded in finding the true signal (X1, X2, X3), with coefficients close to the real values. However, it also kept several noise predictors (false positives), showing that the chosen $\lambda = 0.05$ may be a bit too weak. A cross-validation step (`cv.glmnet()`) is usually recommended to automatically select the optimal λ, balancing prediction accuracy with sparsity.

The next step entails systematic hyperparameter tuning under rigorous validation protocols, a process that mitigates overfitting and enhances predictive performance in finance-focused machine learning workflows.

5.1.8 Optimizing Model Performance

Hyperparameter tuning refines a predictive model from a baseline into a calibrated instrument. Unlike parameters estimated from data, hyper-parameters govern the learning process itself—for instance, the architecture of a neural network, the depth of a decision tree, or the regularization strength in regression. Selecting them well can spell the difference between mediocre outcomes and robust financial forecasts.

A common method is *grid search*, which systematically tests all combinations within specified parameter ranges using cross-validation to estimate out-of-sample accuracy. We will use `caret::train()` with `trainControl()` for the grid search:

Listing 5.1.12: Random Forest configuration via grid search with cross-validation.

```
# ---- Packages ----
library(caret)
library(ranger)

set.seed(42)

# ---- Synthetic classification data (20 features) ----
n <- 1000; p <- 20
X <- replicate(p, rnorm(n))
# create a signal from the first 3 features + noise
lin <- 3*X[,1] - 2*X[,2] + 1.2*X[,3] + rnorm(n, sd = 1)
y <- factor(ifelse(lin > 0, "Class1", "Class2"))

df <- as.data.frame(X)
names(df) <- paste0("X", 1:p)
df$y <- y

# ---- CV setup ----
ctrl <- trainControl(method = "cv", number = 5)

# ---- Grid (like sklearn) ----
# mtry must be <= number of predictors (here p = 20)
tune_grid <- expand.grid(
  mtry = c(5, 10, 20),
  splitrule = "gini",
  min.node.size = c(1, 2, 4)  # ~ min_samples_leaf
)

n_trees_vec <- c(50, 100, 200)  # ~ n_estimators
max_depth_vec <- c(5, 10, 20)   # ~ max_depth

models <- list()
results <- list()
k <- 1

for (nt in n_trees_vec) {
  for (md in max_depth_vec) {
    m <- train(
      y ~ ., data = df,
      method = "ranger",
      trControl = ctrl,
      tuneGrid = tune_grid,
      metric = "Accuracy",
      num.trees = nt,          # n_estimators
      max.depth = md,          # max_depth
      importance = "impurity" # optional
    )
```

```
    models[[k]] <- m
    r <- m$results
    r$num.trees <- nt
    r$max.depth <- md
    results[[k]] <- r
    k <- k + 1
  }
}

all_res <- do.call(rbind, results)

# ---- Best setting (highest CV Accuracy) ----
best_idx <- which.max(all_res$Accuracy)
best <- all_res[best_idx, ]

cat("Best parameters:\n")
print(list(
  n_estimators      = best$num.trees,
  max_depth         = best$max.depth,
  mtry              = best$mtry,
  min_samples_leaf  = best$min.node.size,
  splitrule         = as.character(best$splitrule)
))
cat("\nBest cross-validation accuracy:", round(best$Accuracy, 4), "\n")
```

```
Best parameters:
$n_estimators
[1] 50
$max_depth
[1] 10
$mtry
[1] 10
$min_samples_leaf
[1] 1
$splitrule
[1] "gini"
Best cross-validation accuracy: 0.892
```

The optimal Random Forest configuration balances moderate tree depth with sufficient ensemble size to achieve nearly 90% cross-validation accuracy. Allowing splits down to single observations, combined with random feature selection at each split, enables the model to capture complex structures while maintaining generalizability.

Randomized search samples parameter combinations from specified distributions, offering a faster, more scalable alternative to exhaustive grid search for large datasets or complex models. For randomized search with cross-validation in **caret**, use `search = "random"` with `tuneLength` specifying how many random sets to try.

Listing 5.1.13: Randomized search with cross-validation.

```
# ---- Packages ----
library(caret)
library(ranger)   # random forest engine
set.seed(42)

# ---- Synthetic classification data ----
n <- 500; p <- 10
X <- replicate(p, rnorm(n))
lin <- 3*X[,1] - 2*X[,2] + 1.2*X[,3] + rnorm(n, sd = 1)
y <- factor(ifelse(lin > 0, "Class1", "Class2"))

df <- as.data.frame(X)
names(df) <- paste0("X", 1:p)
df$y <- y

# ---- Train/test split ----
idx <- createDataPartition(df$y, p = 0.8, list = FALSE)
train <- df[idx,]
test  <- df[-idx,]

# ---- Randomized search setup ----
ctrl <- trainControl(method = "cv", number = 5, search = "random")

# Tune grid (ranges analogous to Python's randint)
rf_random <- train(
  y ~ .,
  data = train,
  method = "ranger",
  trControl = ctrl,
  tuneLength = 20,       # number of random combinations
  metric = "Accuracy",
  importance = "impurity"
)

# ---- Results ----
print(rf_random)
plot(rf_random)
```

```
Random Forest

401 samples
```

```
 10 predictor
  2 classes: 'Class1', 'Class2'

No pre-processing
Resampling: Cross-Validated (5 fold)
Summary of sample sizes: 321, 321, 321, 321, 320
Resampling results across tuning parameters:

  min.node.size  mtry  splitrule   Accuracy   Kappa
   2              4    gini        0.8354012  0.6708682
   2              7    extratrees  0.8628704  0.7257426
   2              8    extratrees  0.8702778  0.7406058
   3              9    gini        0.8428704  0.6858234
   5              9    extratrees  0.8552160  0.7104557
   8              8    extratrees  0.8677778  0.7355495
  10              1    extratrees  0.8354321  0.6706908
  10              3    extratrees  0.8653395  0.7307137
  10              4    gini        0.8354321  0.6708895
  10              9    extratrees  0.8602469  0.7203936
  11              2    extratrees  0.8578395  0.7155415
  13              2    extratrees  0.8528395  0.7057204
  13              4    extratrees  0.8703086  0.7405722
  13              9    extratrees  0.8602469  0.7204491
  15              2    extratrees  0.8503704  0.7006817
  15              9    extratrees  0.8603086  0.7205022
  17             10    extratrees  0.8577778  0.7155619
  18              8    gini        0.8378704  0.6756987
  18             10    gini        0.8379012  0.6756504
  20              8    gini        0.8353704  0.6708110

 Accuracy was used to select the optimal model using the largest
 value. The final values used for the model were mtry = 4,
 splitrule = extratrees and min.node.size = 13.
```

The Random Forest achieved strong performance, with cross-validated accuracy ranging from 0.83 to 0.87 and Kappa from 0.67 to 0.74. Models using the *extratrees* split rule consistently outperformed those using *gini*. The optimal configuration was `mtry = 4`, `splitrule = extratrees`, and `min.node.size = 13`, yielding the highest accuracy (0.87) and Kappa (0.74), indicating robust and reliable classification (Figure 5.1-5).

Randomized search balances thoroughness and efficiency but still requires numerous model fits, making it time-intensive without parallel processing. Advanced optimization methods, including Bayesian optimization and genetic algorithms, improve on brute-force search. Bayesian optimization (Listing 5.1.14) constructs probabilistic models linking hyper-parameters to performance

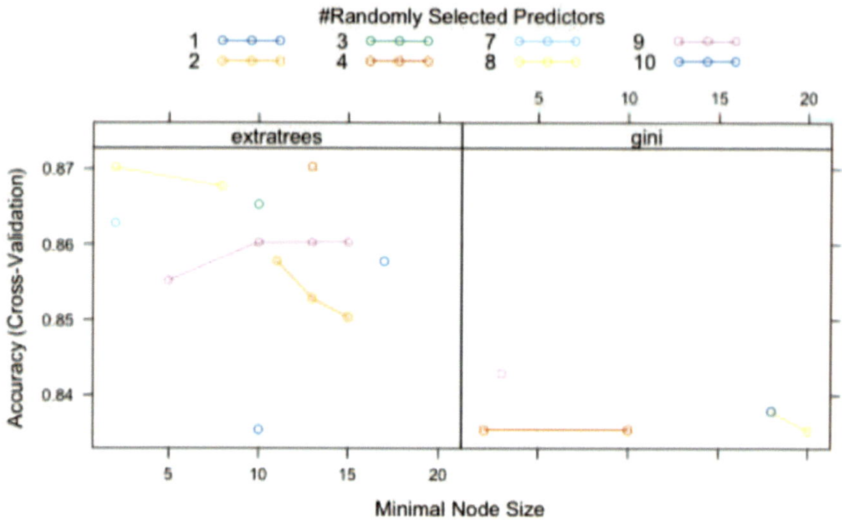

FIGURE 5.1-5: Cross-validated accuracy of Random Forest models across tuning parameters. Models using the *extratrees* split rule consistently outperform *gini*, achieving stable performance above 0.85 and peaking near 0.87 at moderate node sizes (∼10–13) with four randomly selected predictors.

metrics and iteratively selects candidates, converging faster than grid or random search.

Listing 5.1.14: Bayesian optimization.

```
# ---- Packages ----
library(lightgbm)
library(ParBayesianOptimization)
library(ggplot2)
library(GGally)

# Prevent Quartz bitmap error on macOS
options(bitmapType = "cairo")

set.seed(42)

# ---- Synthetic classification data ----
n <- 1000; p <- 20
X <- replicate(p, rnorm(n))
# make task harder (less separable, more noise)
lin <- 1.5*X[,1] - 1*X[,2] + 0.8*X[,3] + rnorm(n, sd = 2)
y <- ifelse(lin > 0, 1, 0)

X_mat <- as.matrix(X)
dtrain <- lgb.Dataset(data = X_mat, label = y)
```

```
# ---- Scoring function with error handling ----
scorer <- function(num_leaves, learning_rate, feature_fraction,
                   bagging_fraction, bagging_freq, min_child_samples) {
  tryCatch({
    params <- list(
      objective = "binary",
      metric = "auc",
      boosting_type = "gbdt",
      verbosity = -1,
      feature_pre_filter = FALSE,
      num_leaves = as.integer(num_leaves),
      learning_rate = learning_rate,
      feature_fraction = feature_fraction,
      bagging_fraction = bagging_fraction,
      bagging_freq = as.integer(bagging_freq),
      min_child_samples = as.integer(min_child_samples)
    )

    cv <- lgb.cv(
      params = params,
      data = dtrain,
      nrounds = 200,
      nfold = 5,
      stratified = TRUE,
      verbose = -1,
      early_stopping_rounds = 20
    )

    best_auc <- max(cv$record_evals$valid$auc$eval)
    list(Score = best_auc, Pred = 0)
  }, error = function(e) {
    message("Skipping invalid config: ", e$message)
    list(Score = runif(1, 0.45, 0.55), Pred = 0) # fallback with
    variance
  })
}

# ---- Parameter search space ----
bounds <- list(
  num_leaves = c(10L, 200L),
  learning_rate = c(0.01, 0.3),
  feature_fraction = c(0.5, 1.0),
  bagging_fraction = c(0.5, 1.0),
  bagging_freq = c(1L, 7L),
  min_child_samples = c(5L, 100L)
)

# ---- Run Bayesian Optimization ----
opt_results <- bayesOpt(
```

```
  FUN = scorer,
  bounds = bounds,
  initPoints = 12,   # must be >= number of parameters (6)
  iters.n = 25,
  acq = "ucb",
  kappa = 2.576
)

cat("\nBest Parameters Found:\n")
print(getBestPars(opt_results))

# ---- Extract results ----
df <- opt_results$scoreSummary
df$Iteration <- seq_len(nrow(df))

# ---- Progress plot ----
p1 <- ggplot(df, aes(x = Iteration, y = Score)) +
  geom_line(color = "steelblue", size = 1) +
  geom_point(color = "darkorange", size = 2) +
  geom_hline(yintercept = max(df$Score), linetype = "dashed", color = "
    red") +
  labs(
    title = "Bayesian Optimization Progress",
    x = "Iteration",
    y = "Validation AUC"
  ) +
  theme_minimal(base_size = 14)

ggsave("bayesopt_progress.png", p1, width = 7, height = 5, dpi = 300)

# ---- Parallel Coordinates Plot of Top Trials ----
df_top <- df[order(-df$Score), ][1:20, ]  # top 20 trials

p2 <- ggparcoord(
  data = df_top,
  columns = which(names(df_top) %in% c("num_leaves","learning_rate",
                                        "feature_fraction","bagging_
    fraction",
                                        "bagging_freq","min_child_
    samples")),
  groupColumn = "Score",
  scale = "uniminmax"
) +
  scale_color_viridis_c() +
  labs(title = "Top 20 Trials: Hyperparameters vs AUC",
       x = "Hyperparameters", y = "Scaled Values") +
  theme_minimal(base_size = 14)

ggsave("bayesopt_parcoords.png", p2, width = 8, height = 5, dpi = 300)
```

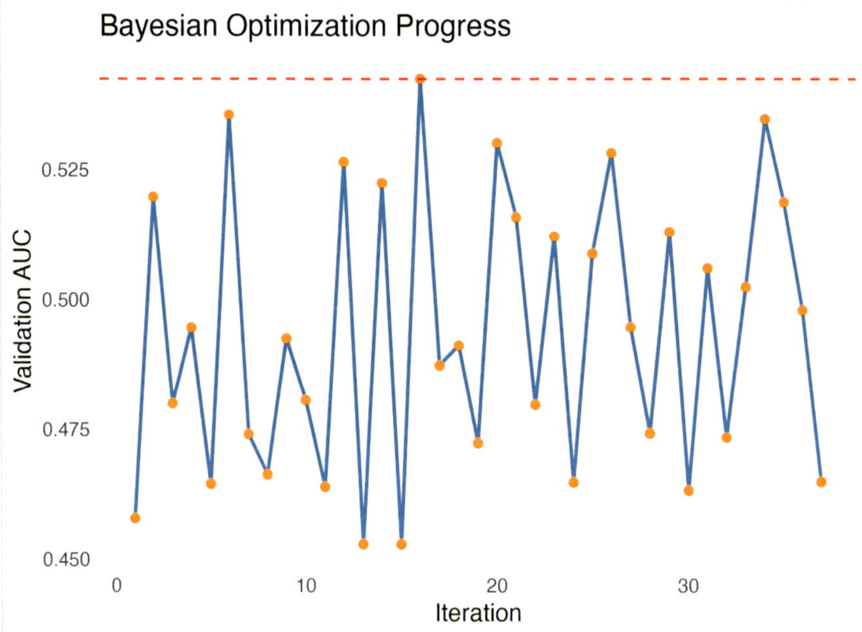

FIGURE 5.1-6: Validation AUC across Bayesian optimization trials. Each orange point represents one trial, while the blue line connects successive results. The red dashed line marks the best observed score.

The optimization progress plot (Figure 5.1-6) illustrates how the validation AUC fluctuated over 35 Bayesian optimization trials. Performance oscillated sharply between weaker and stronger runs, with scores ranging from roughly 0.45 to just under 0.54. This volatility reflects the exploratory strategy of Bayesian optimization, which deliberately samples diverse parameter sets to balance exploration with exploitation. Despite this variability, the algorithm successfully identified higher-performing configurations, as indicated by several peaks above 0.53 and the best result approaching 0.54.

This parallel coordinates plot (Figure 5.1-7) visualizes the top 20 hyperparameter configurations from the Bayesian optimization process, with line color indicating validation AUC. Clear trends emerge: higher-scoring trials (yellow-green) are associated with moderate learning rates, larger numbers of leaves, and smaller minimum child sample sizes, all of which promote greater tree flexibility. Stronger AUCs also cluster at higher feature and bagging fractions, suggesting that models trained on nearly all features and samples performed more robustly. In contrast, weaker configurations (purple) often involved extreme or restrictive values across these parameters.

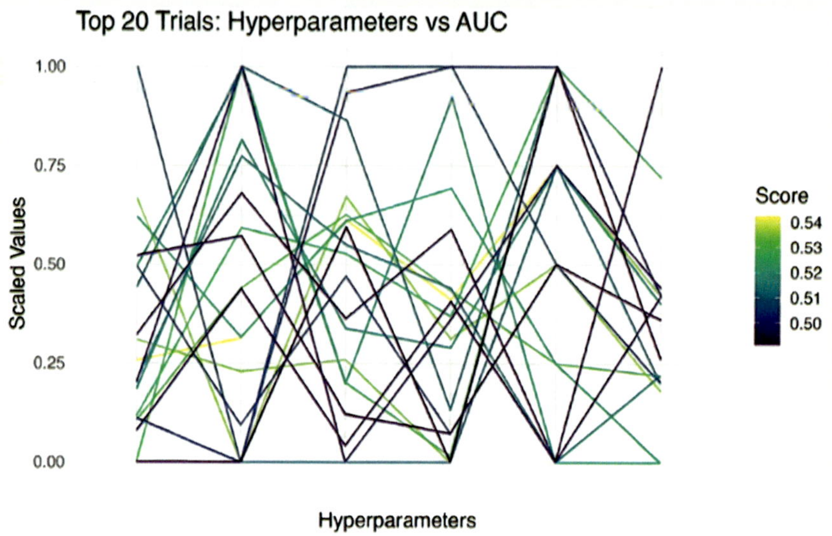

FIGURE 5.1-7: Parallel coordinates plot of the top 20 LightGBM trials. Each line represents one hyperparameter set, with color encoding AUC. Better-performing runs favored moderate learning rates, more leaves, higher feature and bagging fractions, and smaller child sample thresholds..

Figure 5.1-8 illustrates the relative influence of six hyperparameters. In permutation importance, a higher absolute value indicates that the model's performance is more sensitive to changes in that specific parameter.

Listing 5.1.15: Hyperparameter importance analysis.

```
# ---- Random forest permutation importance ----
library(ranger)
library(ggplot2)

df <- opt_results$scoreSummary
df_imp <- df[, c("Score", "num_leaves","learning_rate",
                 "feature_fraction","bagging_fraction",
                 "bagging_freq","min_child_samples")]

# Train random forest surrogate model
rf_model <- ranger(
  Score ~ .,
  data = df_imp,
  importance = "permutation",
  num.trees = 500
)

# Extract and format importance
```

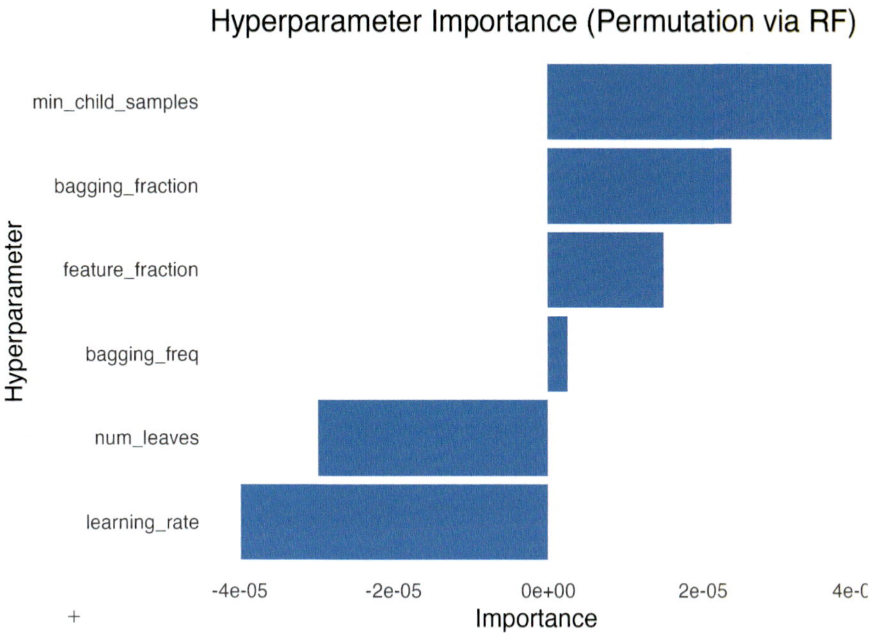

FIGURE 5.1-8: Permutation-based hyperparameter importance estimated via a random forest surrogate. Minimum child samples and bagging fraction emerged as the strongest drivers of AUC, while learning rate and number of leaves had weaker or negligible influence in this dataset.

```
imp_df <- data.frame(
  feature = names(rf_model$variable.importance),
  importance = rf_model$variable.importance
)

# Plot importance
p3 <- ggplot(imp_df, aes(x = reorder(feature, importance), y =
    importance)) +
  geom_col(fill = "steelblue") +
  coord_flip() +
  labs(
    title = "Hyperparameter Importance (Permutation via RF)",
    x = "Hyperparameter",
    y = "Importance"
  ) +
  theme_minimal(base_size = 14)

# Save instead of relying on device
ggsave("bayesopt_importance.png", p3, width = 7, height = 5, dpi = 300)
```

```
cat("\nPermutation importance saved as bayesopt_importance.png\n")
print(imp_df[order(-imp_df$importance), ])
```

This hyperparameter importance analysis, derived from a random forest surrogate with permutation-based scoring (Listing 5.1.15), highlights which parameters most strongly influenced AUC during Bayesian optimization (Figure 5.1-8). The results show that *minimum child samples* and *bagging fraction* had the largest positive impact, suggesting that allowing smaller leaf sizes and training on more data per round contributed most to higher validation scores. Feature fraction also played a meaningful role, while bagging frequency showed only marginal influence. Interestingly, learning rate and number of leaves, though theoretically important, exhibited little or even negative importance in this noisy setting, likely because their effects overlapped with other parameters or because strong performance depended more on sampling strategies than tree depth or learning rate.

Hyperparameter importance analysis exploits information from earlier trials to guide subsequent searches, enabling faster convergence to optimal hyperparameters without exhaustive enumeration.

In finance, cross-validation must preserve temporal order to prevent look-ahead bias. Unlike standard k-fold, or k-fold cross-validation resampling, time-series methods such as **caret** (`createTimeSlices()`) partition data sequentially to respect time dependence.

Listing 5.1.16: **caret**, `createTimeSlices()`.

```
# ---- Packages ----
library(caret)

# ---- Example data ----
set.seed(42)
X <- 1:100   # synthetic time series index

# ---- Time series cross-validation ----
time_slices <- createTimeSlices(1:length(X),
                                initialWindow = 60,   # size of first training set
                                horizon = 10,         # size of test set
                                fixedWindow = FALSE)  # expanding window

# ---- Print splits ----
for (i in seq_along(time_slices$train)) {
  train_index <- time_slices$train[[i]]
  test_index  <- time_slices$test[[i]]
  cat("TRAIN:", train_index, "\n")
  cat("TEST :", test_index, "\n\n")
}
```

TABLE 5.8: Cross-validation time splits. Each fold trains on consecutive indices and tests on the following block.

Fold	Train Indices	Test Indices
1	1–60	61–70
2	1–61	62–71
...		
3	1–90	91–100

Embedding these splits in grid or randomized searches enforces a validation scheme consistent with trading, where only past information can inform future forecasts.

Hyperparameter tuning is deeply intertwined with feature engineering and selection, since the quality of input data often determines how much tuning is necessary in the first place. Rather than rushing into algorithmic optimization, careful preparation of features can reduce the need for complex models or exhaustive searches. When tuning is required, broad parameter grids provide an efficient way to explore the space, with ranges gradually narrowed as performance patterns emerge. This staged refinement, sometimes called *successive halving*, directs resources toward promising regions while discarding unlikely configurations. The process can be accelerated through parallelization, where multiple trials run simultaneously using tools such cloud-based platforms—an approach that becomes essential for large-scale financial data or computationally intensive architectures like deep neural networks. Overfitting remains a persistent risk during tuning, but it can be managed by incorporating early stopping, which halts training once validation performance stabilizes. This not only conserves computation but also strengthens generalization, ensuring that models perform reliably on new data.

In summary, hyperparameter tuning in financial machine learning uniquely blends art and science. It necessitates a disciplined approach to experimentation, guided by deep domain expertise and validation methods that accurately mirror real market conditions, rather than relying on idealized assumptions. When done meticulously, tuning elevates models from mere approximations to powerful, deployable tools—offering a vital edge that can be game-changing when real capital is at stake.

5.1.9 Ensemble Learning for Financial Prediction

Ensemble methods effectively harness the strengths of multiple models to produce reliable and robust financial predictions. By integrating various models rather than relying on a single algorithm, these methods significantly reduce the risk associated with flawed predictors. This approach mirrors portfolio diversification, where spreading capital across assets mitigates financial risk, and similarly stabilizes outcomes by minimizing prediction errors.

Random forests exemplify this concept, aggregating outputs from numerous decision trees trained on different data subsets. This collective decision-making balances biases and errors, ensuring that the overall forecast remains stable, even when individual trees may overfit due to noisy historical data.

Boosting is another powerful strategy where models are built sequentially, with each new model learning from the mistakes of the previous one. Gradient Boosting Machines (GBMs) are a prime example, widely used in finance for their accuracy and flexibility. They excel at uncovering subtle nonlinear relationships, especially in complex tasks such as predicting loan defaults.

TABLE 5.9: Comparison of two leading ensemble methods in finance. Random forests emphasize stability through aggregation, while GBMs emphasize accuracy and flexibility through sequential error correction.

Method	Mechanism	Financial Applications
Random Forests	Aggregate decisions from many trees trained on different data subsets and features; errors average out across the ensemble.	Stable forecasting of stock returns, credit default risk, volatility estimation.
Gradient Boosting Machines (GBMs)	Build trees sequentially, with each iteration correcting predecessor errors; emphasizes hard-to-predict cases.	Loan default prediction, fraud detection, capturing nonlinear market patterns.

To demonstrate boosting in practice, we present a simple R example that applies the *AdaBoost classifier* to a simulated financial dataset.

Listing 5.1.17: AdaBoost classifier example.

```
# ---- Packages ----
library(adabag)       # For AdaBoost
library(rpart)        # For decision trees
library(caret)        # For train/test split and evaluation

set.seed(42)

# ---- Generate synthetic financial dataset ----
# 1000 samples, 20 predictors, with signal in first 10
n <- 1000
p <- 20
X <- replicate(p, rnorm(n))
signal <- 2*X[,1] - 1.5*X[,2] + 1.2*X[,3] - X[,4] + 0.5*X[,5] + rnorm(n
    )
y <- factor(ifelse(signal > 0, "Class1", "Class2"))

df <- as.data.frame(X)
names(df) <- paste0("X", 1:p)
df$y <- y

# ---- Train/test split ----
train_index <- createDataPartition(df$y, p = 0.7, list = FALSE)
```

```r
train_data <- df[train_index, ]
test_data  <- df[-train_index, ]

# ---- Base estimator: shallow decision tree ----
# (equivalent to max_depth = 3)
rpart_control <- rpart.control(maxdepth = 3)

# ---- Initialize AdaBoost with 50 weak learners ----
ada_model <- boosting(
  y ~ .,
  data = train_data,
  mfinal = 50,          # number of weak learners
  control = rpart_control,
  coeflearn = "Zhu"     # common choice for AdaBoost
)

# ---- Predictions ----
ada_pred <- predict(ada_model, newdata = test_data)

# ---- Classification report ----
conf_mat <- confusionMatrix(
  factor(ada_pred$class, levels = levels(test_data$y)),
  test_data$y
)

print(conf_mat)
```

```
Confusion Matrix and Statistics
          Reference
Prediction Class1 Class2
    Class1    127     17
    Class2     22    133
               Accuracy : 0.8696
                 95% CI : (0.826, 0.9056)
    No Information Rate : 0.5017
    P-Value [Acc > NIR] : <2e-16

                  Kappa : 0.7391

 Mcnemar's Test P-Value : 0.5218

            Sensitivity : 0.8523
            Specificity : 0.8867
         Pos Pred Value : 0.8819
         Neg Pred Value : 0.8581
             Prevalence : 0.4983
```

```
         Detection Rate : 0.4247
   Detection Prevalence : 0.4816
      Balanced Accuracy : 0.8695
       'Positive' Class : Class1
```

This example (Listing 5.1.17) illustrates how a sequence of weak learners—here, shallow trees—can be combined into a strong classifier capable of capturing complex market patterns.

Stacking in Ensemble Learning

Stacking extends ensemble learning by combining the predictions of diverse base models through a meta-learner. For example, outputs from logistic regression, support vector machines, and gradient boosting models can be integrated into a single predictive framework. By exploiting the complementary strengths of these algorithms, stacking often delivers superior performance in applications such as stock trend classification or fraud detection, where capturing multiple facets of financial data is essential. Table 5.10 summarizes the key contrasts among bagging, boosting, and stacking in financial machine learning.

Listing 5.1.18: **lightgbm**.

```r
# ---- Packages ----
library(lightgbm)
library(Matrix)
library(pROC)

set.seed(42)

# ---- Synthetic data (replace with your X_train, y_train, X_test, y_
    test) ----
n <- 1000; p <- 10
X <- matrix(rnorm(n * p), nrow = n, ncol = p)
y <- rbinom(n, 1, 0.5)

# Train/test split
train_idx <- sample(1:n, size = 0.7 * n)
X_train <- X[train_idx, ]
y_train <- y[train_idx]
X_test  <- X[-train_idx, ]
y_test  <- y[-train_idx]

# Convert to lgb.Dataset
train_data <- lgb.Dataset(data = X_train, label = y_train)
valid_data <- lgb.Dataset(data = X_test,  label = y_test)

```

```
# ---- Parameters ----
params <- list(
  objective = "binary",
  metric = "auc",
  boosting_type = "gbdt",
  learning_rate = 0.05,
  num_leaves = 31,
  verbose = -1
)

# ---- Train model ----
model <- lgb.train(
  params = params,
  data = train_data,
  nrounds = 1000,
  valids = list(valid = valid_data),
  early_stopping_rounds = 50
)

# ---- Predict and evaluate ----
y_prob <- predict(model, X_test)
roc_obj <- roc(y_test, y_prob)
auc_score <- auc(roc_obj)

cat(sprintf("Validation AUC: %.4f\n", auc_score))
```

```
Validation AUC: 0.5095
```

The validation AUC of 0.5095 indicates that the model performs only marginally better than random guessing. Since an AUC of 0.5 represents no discriminative power at all, this result suggests that the model has failed to capture meaningful patterns in the data and is unable to reliably distinguish between the two classes. Such near-random performance often points to issues such as weak or uninformative features, poor model specification, or potential data imbalance that masks the underlying signal. In its current form, the model provides little predictive value, highlighting the need for further refinement through improved feature engineering, alternative algorithms, or more careful tuning of hyperparameters (Tables 5.10 and 5.11).

Nevertheless, ensembles are not cure-alls. Their use in finance demands cross-validation schemes that account for temporal dependencies in time series, ensuring that performance estimates are not distorted by look-ahead bias.

Integrating ensembling with feature selection enhances performance by reducing dimensionality, concentrating on informative predictors, and minimizing training overhead. When paired with ensemble classifiers, Recursive

TABLE 5.10: Comparison of ensemble learning methods in financial machine learning. Bagging focuses on variance reduction, boosting targets bias reduction, and stacking integrates diverse learners to capture multiple data perspectives.

Method	Core Idea	Financial Applications
Bagging	Parallel combination of models trained on bootstrap samples; reduces variance.	Risk forecasting, volatility estimation, portfolio risk attribution.
Boosting	Sequential combination of weak learners, each correcting predecessor errors; reduces bias.	Credit risk modeling, options pricing anomalies, event-driven trading.
Stacking	Combines heterogeneous models through a meta-learner to exploit complementary strengths.	Stock trend classification, fraud detection, sentiment-driven trading signals.

TABLE 5.11: Diagnostics and recommended next steps for addressing near-random validation AUC results.

Category	Checklist and Actions
Class Balance	Inspect distribution of labels; apply stratified resampling, SMOTE, or class weights if imbalance is severe.
Baseline Models	Compare against trivial baselines (random classifier, class priors, simple logistic regression). Ensure the model beats these.
Feature Signal	Examine univariate feature–target relationships; remove irrelevant predictors or engineer stronger domain features.
Data Integrity	Check for label leakage, misalignment between features and targets, and consistency across training/validation splits.
Overfitting vs. Underfitting	Compare training vs. validation AUC. High training but low validation suggests overfitting; both low suggests weak model specification.
Model Alternatives	Experiment with different families (tree-based vs. linear models, boosting vs. regularized regression).
Hyperparameter Tuning	Use wider or more systematic search (grid, random, Bayesian optimization) to avoid poor defaults.

Feature Elimination (RFE) systematically discards irrelevant variables to yield a more efficient model. The following example uses **caret**, which provides recursive feature elimination with cross-validation via `rfe()` and `rfeControl`.

Listing 5.1.19: Recursive Feature Elimination (RFE) with AdaBoost using **caret** and **adabag**.

```
# ---- Packages ----
library(caret)
library(adabag)

set.seed(42)

# ---- Example synthetic data ----
n <- 200; p <- 10
X <- replicate(p, rnorm(n))
```

```
y <- factor(ifelse(2*X[,1] - 1.5*X[,2] + rnorm(n) > 0, "Class1", "
    Class2"))
df <- as.data.frame(X)
names(df) <- paste0("X", 1:p)
df$y <- y

# ---- Custom AdaBoost wrapper for RFE ----
ada_fit <- function(x, y, first, last, ...) {
  dat <- data.frame(x, y = y)
  adabag::boosting(y ~ ., data = dat, mfinal = 20)
}

ada_pred <- function(object, x) {
  predict(object, newdata = x)$class
}

ada_rank <- function(object, x, y) {
  # importance is a named numeric vector
  imp <- object$importance
  out <- data.frame(var = names(imp), Overall = as.numeric(imp))
  out
}

# ---- RFE Control ----
ctrl <- rfeControl(
  functions = caretFuncs,
  method = "cv",
  number = 5
)

# Override caretFuncs with our custom AdaBoost versions
ctrl$functions$fit  <- ada_fit
ctrl$functions$pred <- ada_pred
ctrl$functions$rank <- ada_rank

# ---- Run RFE ----
set.seed(42)
rfe_fit <- rfe(
  x = df[, -ncol(df)],
  y = df$y,
  sizes = 1:p,
  rfeControl = ctrl
)

# ---- Results ----
print(rfe_fit)
cat("Optimal number of features:", rfe_fit$optsize, "\n")
cat("Selected predictors:", predictors(rfe_fit), "\n")
```

TABLE 5.12: Variable selection results with corresponding accuracy, Kappa, and standard deviations. The best subset (4 variables) is marked with an asterisk.

Variables	Accuracy	Kappa	AccuracySD	KappaSD	Selected
1	0.710	0.42	0.06519	0.13038	
2	0.700	0.40	0.06374	0.12748	
3	0.865	0.73	0.03354	0.06708	
4	0.865	0.73	0.04183	0.08367	*
5	0.835	0.67	0.04183	0.08367	
6	0.845	0.69	0.03708	0.07416	
7	0.830	0.66	0.04108	0.08216	
8	0.820	0.64	0.04472	0.08944	
9	0.850	0.70	0.03953	0.07906	
10	0.850	0.70	0.05303	0.10607	

```
# ---- Plot performance ----
plot(rfe_fit, type = c("g", "o"))
```

```
Recursive feature selection
Outer resampling method: Cross-Validated (5 fold)
Resampling performance over subset size:

The top 4 variables (out of 4):
   X1, X2, X3, X8

Optimal number of features: 4
Selected predictors: X1 X2 X3 X8
```

This RFE run used 5-fold cross-validation to evaluate predictive performance across different subset sizes. The procedure identified that the model performs best when four predictors are retained. Specifically, variables X1, X2, X3, and X8 were consistently ranked as the most informative features.

The result implies that these four variables capture the essential signal in the data, while additional predictors do not materially improve generalization. By focusing on this reduced set, the model gains interpretability and parsimony without sacrificing accuracy, suggesting that noise or redundancy in the other variables can be safely discarded.

The analysis (Table 5.12) confirms that a compact model using only 4 features is sufficient to capture the predictive signal. Beyond this point, additional variables add little value and may even dilute generalization. This highlights the importance of feature selection for balancing accuracy, stability, and interpretability.

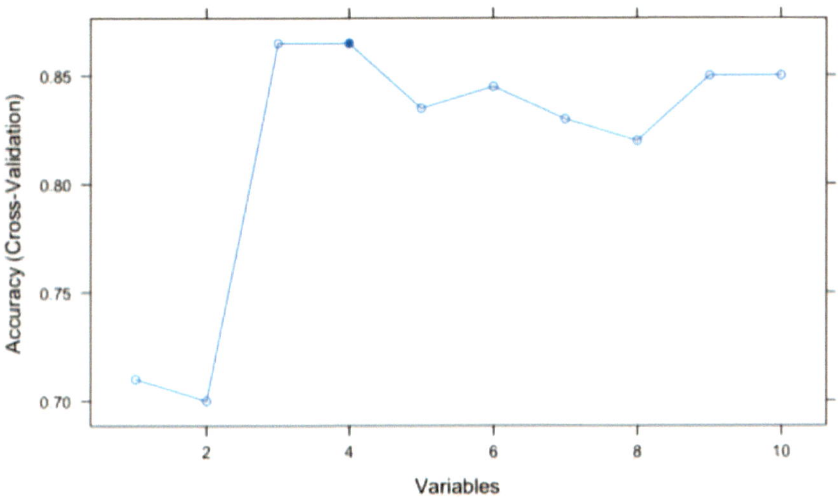

FIGURE 5.1-9: Cross-validated accuracy across different numbers of variables.

The plot shows cross-validated accuracy as a function of the number of variables included in the model. Accuracy improves sharply when moving from 2 to 3 variables, reaching its peak (≈ 0.865) at 3 to 4 variables. After that point, adding more predictors does not enhance performance and in some cases slightly reduces accuracy, indicating diminishing returns. The optimal model complexity lies at four variables, consistent with the table where this subset was selected as best. This suggests that parsimony—using fewer but more informative predictors—outperforms models with redundant or noisy features.

Ensemble methods ensure that only predictive features inform model decisions, a major advantage when dealing with high-dimensional financial data. Beyond classification, ensembles extend naturally to regression problems such as forecasting asset returns or estimating risk, with gradient boosting and related methods adapting well to continuous targets.

At their core, ensembles balance bias and variance by pooling diverse hypotheses while maintaining sufficient flexibility to capture market structure. In predictive modeling, the principle of diversification applies just as in portfolio management: relying on multiple models reduces the risk of overfitting to noise.

Financial institutions employ ensemble frameworks across domains—credit scoring, fraud detection, and algorithmic trading—because they consistently improve predictive stability in noisy, uncertain environments. Mastery of these techniques equips practitioners with rigor and adaptability, traits essential in evolving financial markets where robustness against volatility is critical.

TABLE 5.13: Comparison of Random Forests and Gradient Boosted Trees in financial modeling. Both belong to the ensemble family but emphasize different aspects of the bias–variance tradeoff.

Method	Random Forests	Gradient Boosted Trees
Bias–Variance Tradeoff	Reduces variance through bagging; stable but may retain bias.	Sequentially reduces bias by correcting prior errors; risk of overfitting if not tuned.
Interpretability	Moderate: feature importance scores and partial dependence plots.	Lower: models are complex and harder to explain without SHAP/advanced tools.
Strengths	Handles noisy predictors well; robust to overfitting; good baseline.	High predictive accuracy; adapts to complex nonlinear structures.
Weaknesses	May underfit subtle patterns; requires many trees for best performance.	Sensitive to hyperparameters; longer training times.
Financial Applications	Credit scoring, anomaly detection, risk modeling.	Return forecasting, volatility estimation, trading signal generation.

5.2 Unsupervised Learning (Clustering)

5.2.1 K-Means

Market segmentation is a vital challenge in finance that enhances decision-making. By identifying distinct groups within large pools of assets or investors, professionals can develop tailored strategies and optimize portfolio management. Clustering, an effective form of unsupervised learning, groups data points by their similarities without relying on predefined labels. This approach clarifies the complexity of financial markets, leading to smarter investment strategies. The following R example illustrates this process using the K-Means algorithm (Listing 5.2.1).

Listing 5.2.1: K-Means clustering.

```
# ---- Packages ----
library(ggplot2)

set.seed(42)

# ---- Sample financial features for 10 stocks ----
data <- data.frame(
  Volatility = c(12, 15, 8, 20, 25, 10, 7, 30, 18, 14),
  AvgReturn  = c(5, 7, 3, 10, 12, 4, 2, 15, 9, 6),
  Volume     = c(100, 150, 80, 200, 250, 90, 70, 300, 180, 140)
)

# ---- Apply KMeans clustering into 3 groups ----
kmeans_result <- kmeans(data, centers = 3, nstart = 25)
```

```
data$Cluster <- factor(kmeans_result$cluster)

# ---- Print results ----
print(data)

# ---- Plot clusters (Volatility vs AvgReturn) ----
ggplot(data, aes(x = Volatility, y = AvgReturn, color = Cluster)) +
  geom_point(size = 3) +
  labs(
    x = "Volatility (%)",
    y = "Average Return (%)",
    title = "Stock Clusters Based on Volatility and Return"
  ) +
  theme_minimal()
```

TABLE 5.14: K-Means clustering results for synthetic financial data with three clusters based on volatility, average return, and volume.

Volatility	AvgReturn	Volume	Cluster
12	5	100	1
15	7	150	3
8	3	80	1
20	10	200	3
25	12	250	2
10	4	90	1
7	2	70	1
30	15	300	2
18	9	180	3
14	6	140	3

Listing 5.2.1 generates synthetic stock features and applies clustering to partition them into three distinct groups. The resulting scatter plot illustrates how securities with comparable risk–return profiles naturally align within the same cluster. Such segmentation provides quantitative analysts with a systematic means of identifying risk concentrations and uncovering potentially undervalued asset classes.

K-Means requires the number of clusters K to be specified in advance, which is often unknown. The elbow method addresses this by plotting explained variance for different values of K and identifying the point where gains diminish. To obtain a numerical estimate of the elbow point instead of just a plot, we can use the **NbClust**.

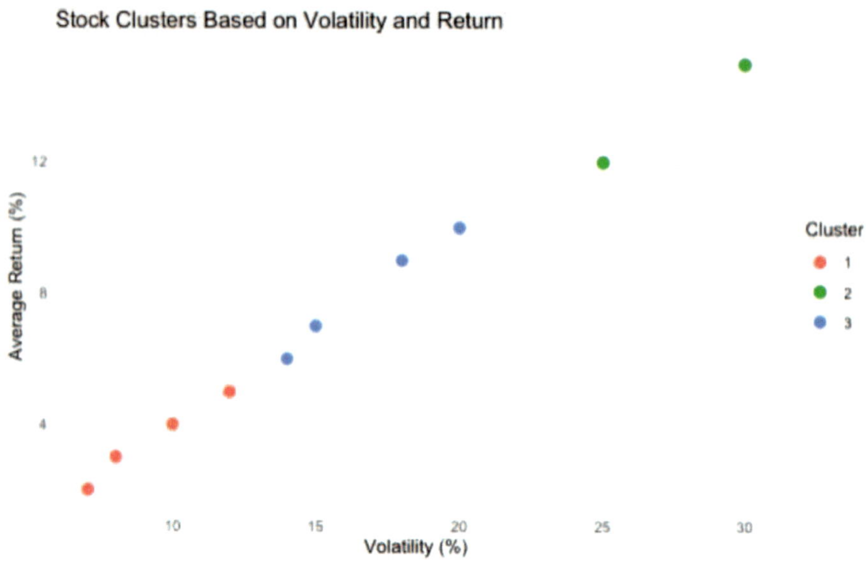

FIGURE 5.2-1: Stock clusters based on volatility and return. Cluster 1 (red) reflects low-risk, low-return stocks; Cluster 3 (blue) represents medium-risk, medium-return profiles; and Cluster 2 (green) captures high-risk, high-return opportunities, illustrating the fundamental tradeoff between risk and reward in financial markets.

Listing 5.2.2: Elbow method to estimate cluster number.

```
# ---- Packages ----
library(ggplot2)
library(factoextra)
library(NbClust)

set.seed(42)

# ---- Example synthetic data ----
n <- 100
df <- data.frame(
  Volatility = runif(n, 5, 30),
  AvgReturn  = runif(n, 2, 15),
  Volume     = runif(n, 100, 1000)
)

# ---- Option 1: Automatic detection with NbClust ----
nb <- NbClust(
  data = df[, c("Volatility", "AvgReturn", "Volume")],
  distance = "euclidean",
  min.nc = 2,
```

```
  max.nc = 9,
  method = "kmeans",
  index = "ch"    # Calinski-Harabasz index
)

cat("Optimal number of clusters (NbClust, CH index):", nb$Best.nc[1], "
    \n")

# ---- Option 2: Elbow method with factoextra ----
fviz_nbclust(
  df[, c("Volatility", "AvgReturn", "Volume")],
  kmeans,
  method = "wss"
) +
  labs(title = "Elbow Method for Optimal k") +
  theme_minimal()
```

```
Optimal number of clusters (NbClust, CH index): 9
```

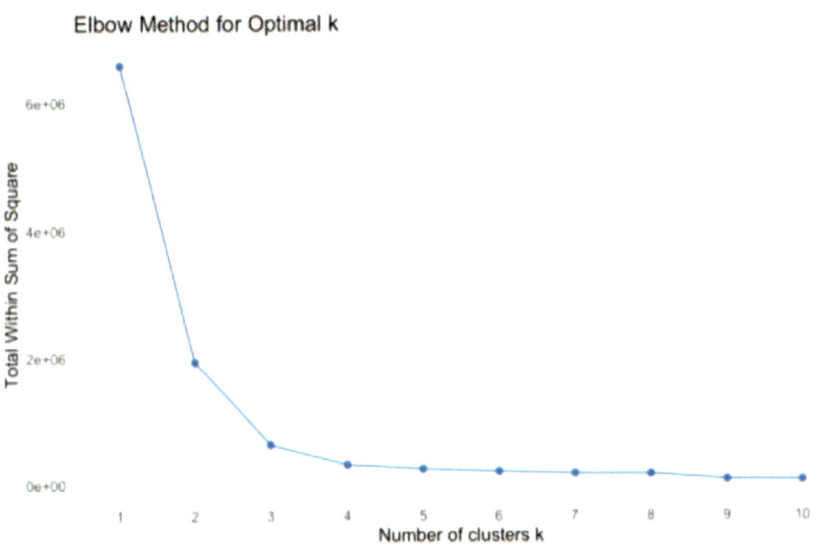

FIGURE 5.2-2: Elbow method plot for estimating the optimal number of clusters. The sharp bend at $k = 3$ indicates a suitable cluster choice, as additional clusters yield only marginal improvements in explained variance.

The elbow method (Figure 5.2-2) suggests that the optimal number of clusters lies around $k = 3$, where the curve flattens and additional clusters contribute little to reducing within-cluster variance. In contrast, **NbClust** (Listing 5.2.2) using the *Calinski–Harabasz (CH)* index recommends $k = 9$. This discrepancy illustrates a common challenge in clustering: different criteria may yield different solutions. The elbow method emphasizes parsimony, preferring fewer clusters with meaningful separation, while the *CH index* favors more clusters when additional partitions still improve model fit. In practice, the choice between $k = 3$ and $k = 9$ depends on domain knowledge and whether interpretability (fewer clusters) or granularity (more clusters) is prioritized.

5.2.2 Hierarchical Clustering

Clustering is a powerful technique that groups similar data points while separating dissimilar ones into distinct clusters. Hierarchical Clustering enhances this by establishing a clear hierarchy of clusters.

Hierarchical Clustering can be approached in two ways: top-down or bottom-up, resulting in two types: Divisive and Agglomerative. Each method offers unique advantages and insights into data relationships. Hierarchical clustering avoids pre-specifying the number of clusters by constructing a dendrogram—a nested tree that merges points or clusters based on their distances. It is well suited to capturing subtle relationships and to exploring data at different levels of granularity.

Agglomerative Clustering with Dendrogram Visualization

Agglomerative clustering follows a bottom-up strategy in which each observation begins as its own cluster, and the algorithm successively merges the most similar clusters until a complete hierarchy is formed (see results Table 5.15).

Listing 5.2.3: Agglomerative Clustering with Dendrogram Visualization.

```r
# ---- Packages ----
library(ggplot2)

set.seed(42)

# ---- Example synthetic data ----
n <- 100
df <- data.frame(
  Volatility = runif(n, 5, 30),
  AvgReturn  = runif(n, 2, 15),
  Volume     = runif(n, 100, 1000)
)

# ---- Distance matrix ----
dist_mat <- dist(df[, c("Volatility", "AvgReturn", "Volume")])
```

TABLE 5.15: Cluster assignments from agglomerative clustering based on volatility, average return, and volume.

ID	Volatility	AvgReturn	Volume	AggloCluster
1	27.87015	10.141189	896.6059	1
2	28.42689	4.823050	565.3999	2
3	12.15349	4.815375	866.7379	1
4	25.76119	7.056285	498.5166	2
5	21.04364	14.251924	242.0921	3
6	17.97740	14.513904	498.0922	2

```
# ---- Hierarchical clustering (Ward's method) ----
hc <- hclust(dist_mat, method = "ward.D2")

# ---- Plot dendrogram ----
plot(hc, labels = FALSE, main = "Dendrogram with 3 Clusters")

# ---- Add colored rectangles for chosen clusters ----
rect.hclust(hc, k = 3, border = c("red", "blue", "darkgreen"))

# ---- Assign clusters ----
df$AggloCluster <- cutree(hc, k = 3)

# ---- Show sample of clustered data ----
print(head(df[, c("Volatility", "AvgReturn", "Volume", "AggloCluster")
    ]))
```

The dendrogram (Figure 5.2-3) shows the results of hierarchical clustering using Ward's minimum variance method (ward.D2). It highlights natural breaks in the data hierarchy that can guide cluster selection. This visualization helps uncover tightly knit sectors or market regimes before formal segmentation occurs.

In applied contexts such as asset allocation, hierarchical clustering is especially useful because it can detect overlapping behaviors that flat clustering methods might miss. For instance, a cluster characterized by moderate volatility but high trading volume may further divide into subgroups, each reflecting distinct sector exposures. A comparison of clustering approaches are tabulated in Table 5.16.

Principal Component Analysis

Financial datasets are often noisy and high-dimensional, hindering clustering performance. Dimensionality reduction with PCA mitigates these issues by capturing most variance in fewer orthogonal features, improving both clarity

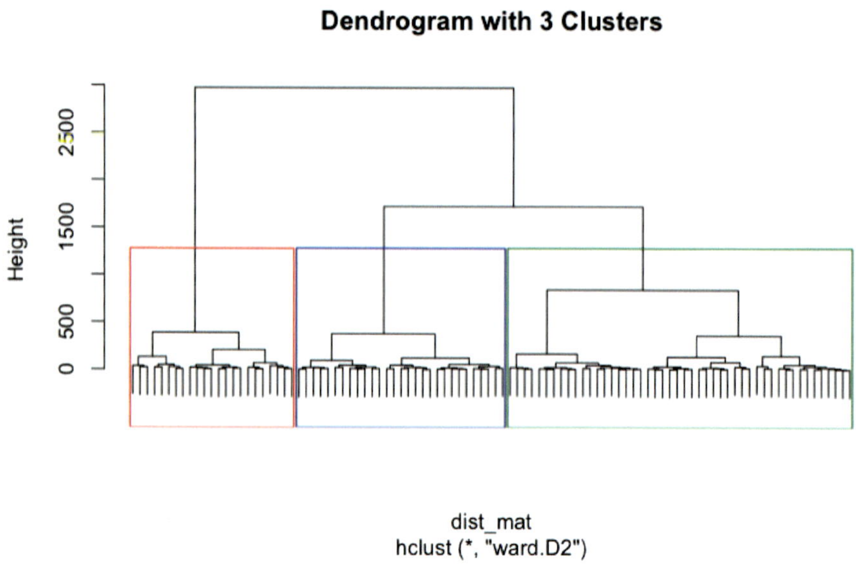

FIGURE 5.2-3: Hierarchical agglomerative clustering dendrogram using Ward's method. The tree is cut at a height yielding three distinct clusters (red, blue, green), each grouping observations with similar characteristics.

TABLE 5.16: Comparison of three clustering approaches in financial applications. Hierarchical clustering (Ward's method) offers interpretability through dendrograms, DBSCAN excels at detecting noise and irregular cluster shapes, and Gaussian Mixture Models (GMM) handle overlapping clusters while quantifying uncertainty.

Method	**Strengths**	**Use Cases in Finance**
Hierarchical Clustering (Ward's Method)	Builds a nested tree (dendrogram) without pre-specifying cluster count; Ward's criterion minimizes within-cluster variance.	Identifying market regimes, grouping assets with similar volatility/return profiles, and exploratory analysis before segmentation.
DBSCAN (Density-Based Spatial Clustering with Noise)	Detects arbitrarily shaped clusters and isolates outliers; does not require pre-defined cluster number.	Fraud detection, anomaly detection in trading data, and uncovering irregular patterns in transaction flows.
Gaussian Mixture Models (GMM)	Probabilistic method modeling data as mixtures of Gaussian distributions; handles overlapping clusters and quantifies uncertainty.	Asset allocation under uncertainty, regime detection when clusters overlap, and modeling return distributions with multiple modes.

and efficiency. Listing 5.2.4 demonstrates the integration of PCA and K-Means clustering, a powerful duo for high-dimensional data analysis.

Listing 5.2.4: Combining PCA with K-Means clustering.

```
# ---- Packages ----
library(ggplot2)
library(cluster)   # for kmeans
library(stats)     # for prcomp

set.seed(42)

# ---- Synthetic data ----
n <- 150
df <- data.frame(
  Volatility = rnorm(n, mean = 20, sd = 5),
  AvgReturn  = rnorm(n, mean = 5, sd = 2),
  Volume     = rnorm(n, mean = 1000, sd = 200)
)

# ---- PCA: reduce from 3D to 2D ----
pca_res <- prcomp(df, scale. = TRUE)
reduced_data <- data.frame(pca_res$x[, 1:2])

# ---- K-Means clustering on reduced data ----
set.seed(42)
kmeans_pca <- kmeans(reduced_data, centers = 3, nstart = 25)
reduced_data$cluster <- factor(kmeans_pca$cluster)

# ---- Visualization ----
ggplot(reduced_data, aes(x = PC1, y = PC2, color = cluster)) +
  geom_point(size = 2) +
  labs(
    title = "Clusters After PCA Dimensionality Reduction",
    x = "Principal Component 1",
    y = "Principal Component 2"
  ) +
  theme_minimal()
```

The scatter plot (Figure 5.2-4) shows the K-Means clustering applied to data reduced with PCA. Reducing the three original features—`Volatility`, `AvgReturn`, and `Volume`—into two principal components captured nearly 88% of the dataset's variance (PC1 $\approx$ 60%, PC2 $\approx$ 28%). Clustering in this reduced space produced three distinct groups with clear centroid differences. Cluster 1 (red) centers around positive PC1 values near +0.8, indicating relatively higher volatility–return combinations. Cluster 2 (green) averages close to +1.1 on PC2, marking a regime that diverges strongly along the secondary variance dimension. Cluster 3 (blue) is anchored at negative PC2 values near –1.0, representing observations with lower scores on the second component.

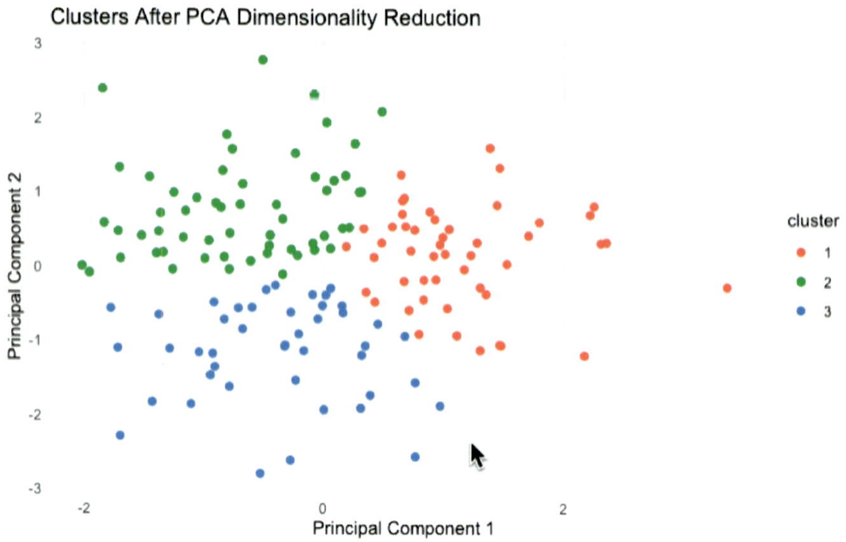

FIGURE 5.2-4: Clusters identified by K-Means after PCA dimensionality reduction.

These quantitative distinctions in PCA space demonstrate how dimensionality reduction clarifies latent structure and enables K-Means to separate regimes that would otherwise be obscured in the raw feature space.

Projecting data into a lower-dimensional space enhances visualization and can expose cluster boundaries that are obscured in the original, correlated feature set. Quantitative analysts tune pre-processing and hyperparameters to derive market segmentations that generalize across regimes; clustering can isolate groups such as small-cap growth or large-cap value, enabling models tailored to each segment's return and risk profile. Clustering enables investors to move beyond intuition toward systematic insight, uncovering the underlying structures beneath market noise.

Techniques for Reducing Dimensionality

Financial datasets may contain hundreds or thousands of variables, from price histories and technical indicators to macroeconomic and sentiment measures. Left unfiltered, such high dimensionality increases computational cost and risk of overfitting, as models capture noise rather than signal. Dimensionality reduction preserves essential information while discarding redundancy and noise.

PCA is widely applied in finance for its simplicity and effectiveness. It transforms correlated variables into a smaller set of uncorrelated components that capture most of the data's variance.

For instance, daily returns from 50 stocks observed over several years can be distilled through PCA into a few principal components that capture most of the variation in market behavior. The following R example illustrates this process.

Listing 5.2.5: PCA applied to daily returns from 50 stocks.

```
# ---- Packages ----
library(stats)       # for prcomp
library(ggplot2)     # for plotting

set.seed(42)

# ---- Simulate daily returns for 50 stocks over 500 days ----
n_days   <- 500
n_stocks <- 50
returns <- matrix(rnorm(n_days * n_stocks, mean = 0, sd = 0.02),
                  nrow = n_days, ncol = n_stocks)
colnames(returns) <- paste0("Stock", 1:n_stocks)

# ---- Standardize returns (mean=0, sd=1) ----
returns_std <- scale(returns)

# ---- Apply PCA retaining 5 components ----
pca_fit <- prcomp(returns_std, center = TRUE, scale. = TRUE)

# ---- Explained variance ratio ----
explained_var <- pca_fit$sdev^2 / sum(pca_fit$sdev^2)
print("Explained variance per component:")
print(round(explained_var[1:5], 4))

# ---- Plot cumulative explained variance ----
cum_var <- cumsum(explained_var)
df_var <- data.frame(
  Component = 1:length(cum_var),
  CumulativeVariance = cum_var
)

ggplot(df_var, aes(x = Component, y = CumulativeVariance)) +
  geom_line() +
  geom_point() +
  labs(
    title = "PCA on Stock Returns",
    x = "Number of Components",
    y = "Cumulative Explained Variance"
  ) +
  theme_minimal()
```

```
[1] "Explained variance per component:"
[1] 0.0330 0.0322 0.0314 0.0300 0.0204
```

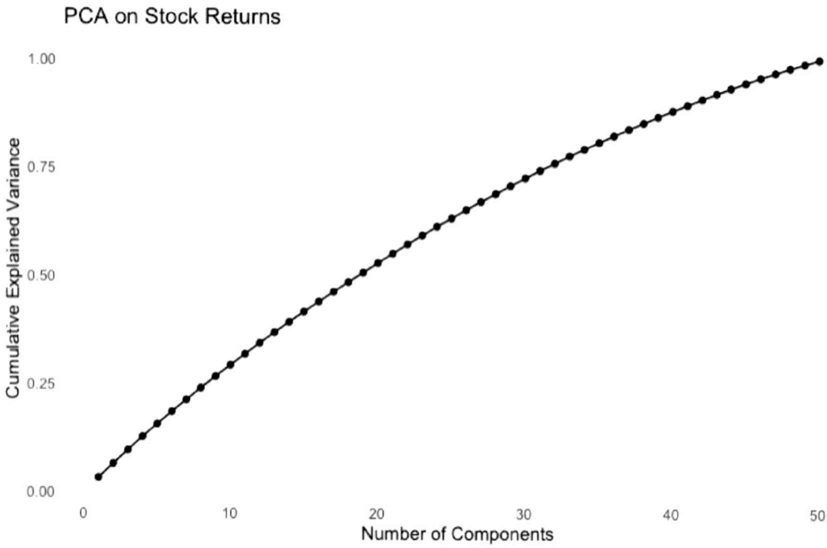

FIGURE 5.2-5: Cumulative explained variance from a principal component analysis (PCA) applied to stock returns.

As displayed in Figure 5.2-5, the curve rises steadily, indicating that each additional component captures incremental variance. The first ≈10 components account for roughly 35–40% of the variance, suggesting strong common factors in stock returns (e.g., broad market and sector effects). By around 25–30 components, more than 70% of the variance is explained, meaning a relatively moderate number of components capture the bulk of systematic variation. To reach nearly full variance explanation (≈100%), all 50 components are needed, which is expected since PCA decomposes the entire covariance structure. Analysts often select a smaller subset of components (e.g., first 5–10) to represent latent risk factors, balancing dimensionality reduction with explanatory power. This avoids overfitting while still capturing the dominant sources of return variation.

Another effective technique is the use of autoencoders—neural networks that learn to compress data into a low-dimensional representation and then reconstruct it with minimal information loss (Listing 5.2.6). In contrast to PCA's strictly linear transformations, autoencoders can capture nonlinear dependencies frequently observed in financial markets, including regime shifts and abrupt volatility spikes. Extensions such as variational autoencoders (VAEs) further enhance this framework by introducing probabilistic interpretations of latent structures.

Listing 5.2.6: Autoencoder using **keras**.

```r
# ---- Packages ----
library(keras)

# ---- Example synthetic data (replace with returns_std) ----
set.seed(42)
n <- 500    # number of samples
p <- 50     # number of features (like stock returns across assets)
returns_std <- matrix(rnorm(n * p), nrow = n, ncol = p)

# ---- Autoencoder parameters ----
input_dim <- ncol(returns_std)
encoding_dim <- 10  # compressed representation size

# ---- Model definition ----
input_layer <- layer_input(shape = input_dim)

encoded <- input_layer %>%
  layer_dense(units = encoding_dim, activation = "relu")

decoded <- encoded %>%
  layer_dense(units = input_dim, activation = "linear")

autoencoder <- keras_model(inputs = input_layer, outputs = decoded)

# ---- Compile ----
autoencoder %>% compile(
  optimizer = "adam",
  loss = "mse"
)

# ---- Fit ----
history <- autoencoder %>% fit(
  x = returns_std,
  y = returns_std,
  epochs = 50,
  batch_size = 32,
  shuffle = TRUE,
  validation_split = 0.2
)
```

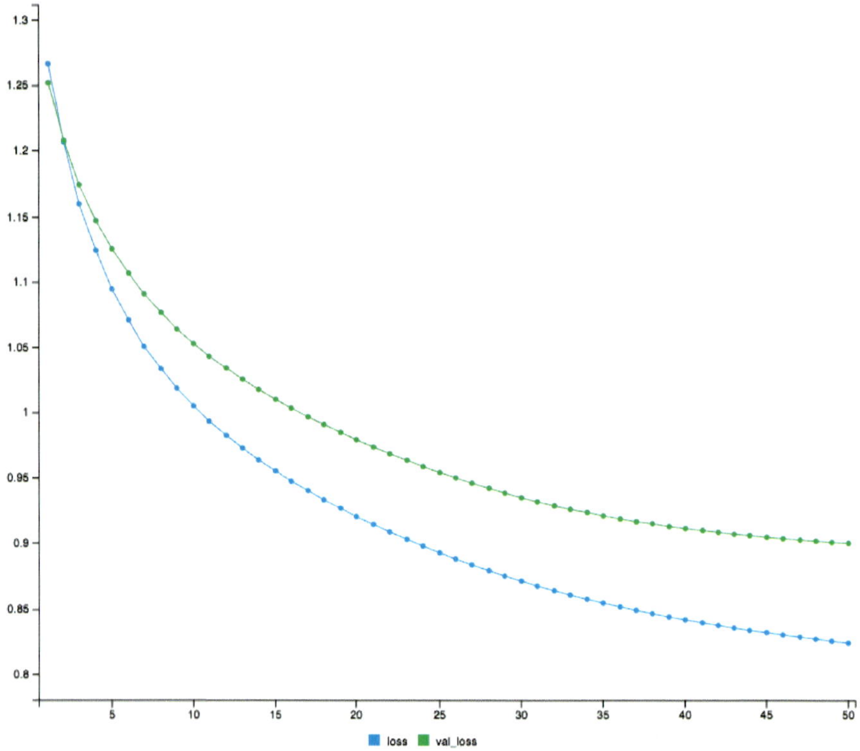

FIGURE 5.2-6: Training and validation loss curves for the autoencoder over 50 epochs.

Figure 5.2-6 shows training and validation loss curves, where:

- **Overall trend**: Both training loss (blue) and validation loss (green) decrease smoothly, confirming that the model is steadily improving its ability to reconstruct the input data.
- **Convergence**: The curves flatten after ≈40 epochs, suggesting the model is approaching its minimum reconstruction error. Additional training would likely bring only marginal gains.
- **Training vs. validation gap**: The validation curve consistently lies above the training curve, which is expected. Importantly, the gap remains stable rather than widening, indicating that the autoencoder is not overfitting but generalizing well.
- **Magnitude of improvement**: Loss drops from ≈1.27 at the start to ≈0.82 (training) and ≈0.90 (validation) by epoch 50, a reduction of about 30–35% in reconstruction error.

Once trained, the encoder produces compressed features that capture non-linear relationships more complex than those detectable by PCA.

Gaussian Mixture Models

Gaussian Mixture Models (GMMs) offer a sophisticated clustering approach that overcomes the limitations of traditional methods like *k-means* and *hierarchical clustering*. Rather than assigning each observation to a single cluster, GMMs recognize that data comes from a mixture of Gaussian distributions, allowing for soft clustering that reveals overlaps and complex structures—especially valuable in finance, where market boundaries are often fluid.

When applied to financial data, GMMs uncover essential patterns, such as stocks that span multiple sectors or markets that shift regimes. Unlike traditional distance-based algorithms, GMMs incorporate covariance, enabling elliptical clusters that better reflect real-world return distributions. This flexibility helps analysts identify asymmetric risk profiles and correlated behaviors, facilitating more informed decision-making in a dynamic financial landscape.

Listing 5.2.7: Cluster assignments and probabilities.

```
# ---- Packages ----
library(tidyverse)
library(mclust)     # for Gaussian Mixture Models

set.seed(42)

# ---- Generate synthetic returns data (100 days x 10 tickers) ----
n_days   <- 100
n_stocks <- 10
tickers  <- paste0("Stock", 1:n_stocks)

# Simulate returns ~ Normal(0, sd=0.02)
returns <- matrix(rnorm(n_days * n_stocks, mean = 0, sd = 0.02),
                  nrow = n_days, ncol = n_stocks,
                  dimnames = list(NULL, tickers))

returns <- as.data.frame(returns)

# ---- Standardize returns (per stock) ----
returns_std <- scale(returns)

# ---- PCA for dimensionality reduction (transpose: stocks x features)
    ----
pca <- prcomp(t(returns_std), center = TRUE, scale. = FALSE)

# Take first 5 principal components
returns_pca <- pca$x[, 1:5]

# ---- Fit Gaussian Mixture Model (4 clusters) ----
```

```
gmm <- Mclust(returns_pca, G = 4)

# ---- Extract cluster labels and probabilities ----
labels <- gmm$classification
probs  <- gmm$z     # posterior probabilities

# ---- Cluster assignments with probabilities ----
cluster_df <- data.frame(
  Ticker = tickers,
  Cluster = labels,
  Prob_Cluster1 = probs[, 1],
  Prob_Cluster2 = probs[, 2],
  Prob_Cluster3 = probs[, 3],
  Prob_Cluster4 = probs[, 4]
)

# ---- Preview sorted by cluster ----
head(cluster_df[order(cluster_df$Cluster), ])
```

TABLE 5.17: Cluster assignments and posterior probabilities from Gaussian Mixture Model with 4 components (synthetic stock returns).

Ticker	Cluster	Prob_Cluster	Prob_Cluster	Prob_Cluster	Prob_Cluster4
Stock3	1	0.91	0.03	0.04	0.02
Stock7	2	0.05	0.85	0.07	0.03
⋮	⋮	⋮	⋮	⋮	⋮

Listing 5.2.8: Visualization of GMM PCA clusters.

```
# ---- Packages ----
library(tidyverse)
library(mclust)    # Gaussian Mixture Models
library(ggrepel)   # for non-overlapping text labels

set.seed(42)

# ---- Generate synthetic returns data (100 days x 10 tickers) ----
n_days   <- 100
n_stocks <- 10
tickers  <- paste0("Stock", 1:n_stocks)

returns <- matrix(rnorm(n_days * n_stocks, mean = 0, sd = 0.02),
                  nrow = n_days, ncol = n_stocks,
                  dimnames = list(NULL, tickers))
returns <- as.data.frame(returns)

# ---- Standardize returns ----
```

```
returns_std <- scale(returns)

# ---- PCA (stocks x features) ----
pca <- prcomp(t(returns_std), center = TRUE, scale. = FALSE)
returns_pca <- pca$x[, 1:5]  # first 5 PCs

# ---- Fit Gaussian Mixture Model (4 clusters) ----
gmm <- Mclust(returns_pca, G = 4)
labels <- gmm$classification

# ---- Data for plotting (first 2 PCs) ----
plot_df <- data.frame(
  Ticker = tickers,
  PC1 = returns_pca[, 1],
  PC2 = returns_pca[, 2],
  Cluster = as.factor(labels)
)

# ---- Visualization with ggplot2 ----
ggplot(plot_df, aes(x = PC1, y = PC2, color = Cluster, label = Ticker))
    +
  geom_point(size = 4, alpha = 0.8) +
  geom_text_repel(show.legend = FALSE, size = 3) +
  theme_minimal(base_size = 14) +
  labs(title = "PCA Cluster Visualization of Synthetic Stocks",
       subtitle = "Gaussian Mixture Model with 4 Components",
       x = "Principal Component 1",
       y = "Principal Component 2") +
  scale_color_brewer(palette = "Set1")
```

As shown in Figure 5.2-7, the 10 stocks split into four distinct clusters, each represented by a different color. The Gaussian Mixture Model captures elliptical groupings rather than strict geometric partitions, allowing stocks with partial overlap to be probabilistically assigned. **Cluster 1 (red)**: Stocks 1 and 4 share very similar return structures, positioned with high PC2 values. **Cluster 2 (blue)**: Stocks 2, 3, 9, and 10 form a spread-out group on the left and lower portions of the PCA space. **Cluster 3 (green)**: Stocks 5 and 6 cluster tightly on the right-hand side, indicating strongly correlated behavior. **Cluster 4 (purple)**: Stocks 7 and 8 sit lower-center, distinct from the others.

These clusters represent latent risk-return structures hidden in the standardized return series. Stocks within a cluster behave similarly, suggesting they may share common risk factors (e.g., exposure to volatility regimes or sector-specific effects). The spread between clusters shows diversification opportunities—e.g., picking stocks from Cluster 1 and Cluster 3 would likely yield less correlated positions.

Unlike hard assignments (k-means), the Gaussian Mixture Model provides probabilities of cluster membership. For example, a stock near the boundary

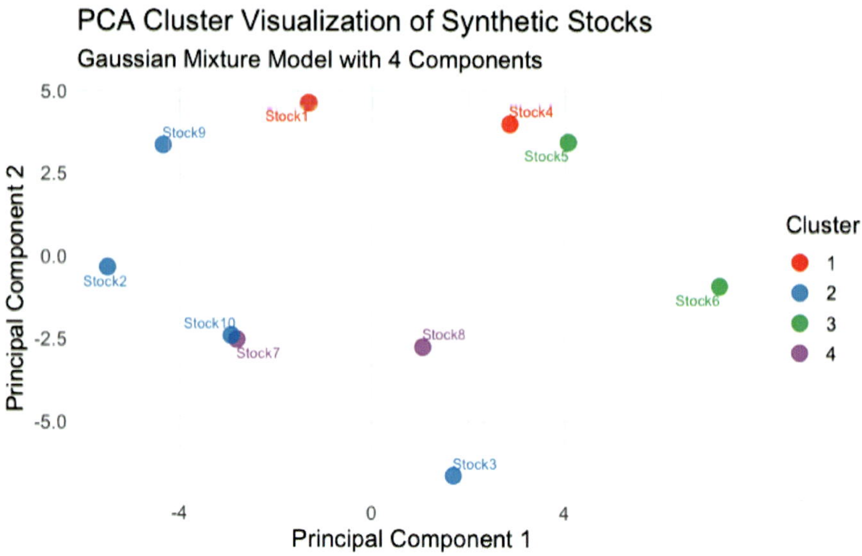

FIGURE 5.2-7: PCA cluster visualization after dimensionality reduction and Gaussian mixture clustering.

of two clusters might have a 60–40 probability split, reflecting partial membership in multiple regimes.

The following code, extends the PCA cluster visualization to include the elliptical contours that represent each Gaussian mixture component's uncertainty region. These ellipses make the latent Gaussian structure visible:

Listing 5.2.9: PCA cluster visualization elliptical contours.

```
# ---- Packages ----
library(tidyverse)
library(mclust)
library(ggrepel)
library(ellipse)    # to draw Gaussian ellipses

set.seed(42)

# ---- Synthetic returns (100 days x 10 tickers) ----
n_days   <- 100
n_stocks <- 10
tickers  <- paste0("Stock", 1:n_stocks)
returns <- matrix(rnorm(n_days * n_stocks, mean = 0, sd = 0.02),
                  nrow = n_days, ncol = n_stocks,
                  dimnames = list(NULL, tickers)) %>% as.data.frame()

# ---- Standardize ----
```

```
returns_std <- scale(returns)

# ---- PCA (stocks x features) ----
pca <- prcomp(t(returns_std), center = TRUE, scale. = FALSE)
returns_pca <- pca$x[, 1:5]

# ---- Fit GMM ----
gmm <- Mclust(returns_pca, G = 4)
labels <- gmm$classification

# ---- Data for plotting ----
plot_df <- data.frame(
  Ticker = tickers,
  PC1 = returns_pca[, 1],
  PC2 = returns_pca[, 2],
  Cluster = as.factor(labels)
)

# ---- Extract ellipse coordinates for each cluster ----
ellipse_df <- do.call(rbind, lapply(1:gmm$G, function(k) {
  mu <- gmm$parameters$mean[1:2, k]
  sigma <- gmm$parameters$variance$sigma[1:2, 1:2, k]
  ell <- as.data.frame(ellipse(sigma, centre = mu, level = 0.95,
    npoints = 200))
  colnames(ell) <- c("x", "y")   # <-- rename here
  ell$Cluster <- as.factor(k)
  ell
}))

# ---- Visualization ----
ggplot(plot_df, aes(x = PC1, y = PC2, color = Cluster)) +
  geom_point(size = 4, alpha = 0.8) +
  geom_text_repel(aes(label = Ticker), show.legend = FALSE, size = 3) +
  geom_path(data = ellipse_df,
            aes(x = x, y = y, color = Cluster),
            inherit.aes = FALSE,
            size = 1, linetype = "dashed") +
  theme_minimal(base_size = 14) +
  labs(title = "PCA Cluster Visualization of Synthetic Stocks",
       subtitle = "Gaussian Mixture Model with 4 Components and 95%
    Confidence Ellipses",
       x = "Principal Component 1",
       y = "Principal Component 2") +
  scale_color_brewer(palette = "Set1")
```

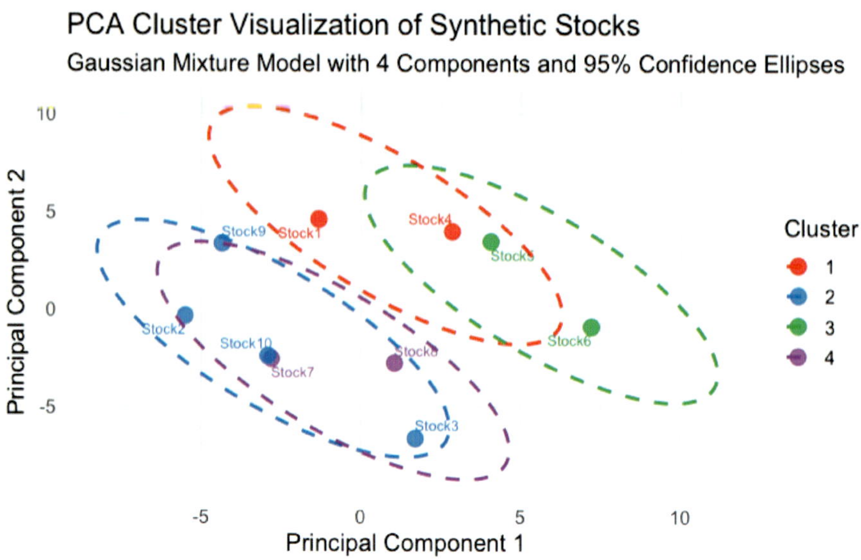

FIGURE 5.2-8: PCA cluster visualization with elliptical contours representing each Gaussian mixture component's uncertainty region.

A central advantage of Gaussian Mixture Models is their ability to assign probabilistic memberships rather than hard labels. Instead of forcing each stock into a single cluster, GMM expresses the degree of belonging across clusters—for instance, a stock may have a 70% probability of being in one group while retaining partial association with others. This nuance is particularly valuable when analyzing firms that straddle multiple business models or exhibit exposure to shifting sectoral cycles. Visualizing clusters provides an intuitive lens for interpreting their underlying structure, revealing how groups form, overlap, and separate within the data space (Figure 5.2-8).

Mastering Gaussian Mixture Models empowers quantitative investors with a powerful tool to reveal hidden asset structures vital for effective diversification and alpha generation in today's intricate markets. This approach acts with surgical precision, setting itself apart from the blunt methods of traditional clustering techniques.

5.3 Deep Learning (Neural Networks)

Neural networks underpin deep learning, a branch of machine learning that has revolutionized pattern recognition and predictive modeling. By emulating the interconnected structure of neurons in the human brain, they can learn complex nonlinear relationships within data. This capability is particularly valuable in financial markets, where hidden patterns often elude linear models. Neural networks are especially effective at capturing subtle dependencies across time and assets, making them a powerful instrument for financial analysis.

5.3.1 Feedforward Neural Network

Take the example of a feedforward neural network used to predict stock price movements (Listing 5.3.1). The input layer may consist of normalized daily returns, trading volume, and selected technical indicators. These inputs are propagated through hidden layers, where weighted sums and activation functions iteratively transform the data, ultimately yielding an output such as the probability that the stock price will increase the next day.

Listing 5.3.1: Feedforward neural network for binary classification.

```
# ---- Packages ----
library(keras)

# ---- Sample dataset: 1000 samples, 10 features ----
set.seed(42)
X_train <- matrix(runif(1000 * 10), nrow = 1000, ncol = 10)
y_train <- sample(0:1, 1000, replace = TRUE)

# ---- Define model ----
model <- keras_model_sequential() %>%
  layer_dense(units = 32, activation = "relu", input_shape = c(10)) %>%
  layer_dense(units = 16, activation = "relu") %>%
  layer_dense(units = 1, activation = "sigmoid")

# ---- Compile model ----
model %>% compile(
  optimizer = "adam",
  loss = "binary_crossentropy",
  metrics = "accuracy"
)

# ---- Fit model ----
history <- model %>% fit(
  X_train, y_train,
  epochs = 10,
  batch_size = 32,
  verbose = 1
)

# ---- Evaluate model ----
model %>% evaluate(X_train, y_train)
```

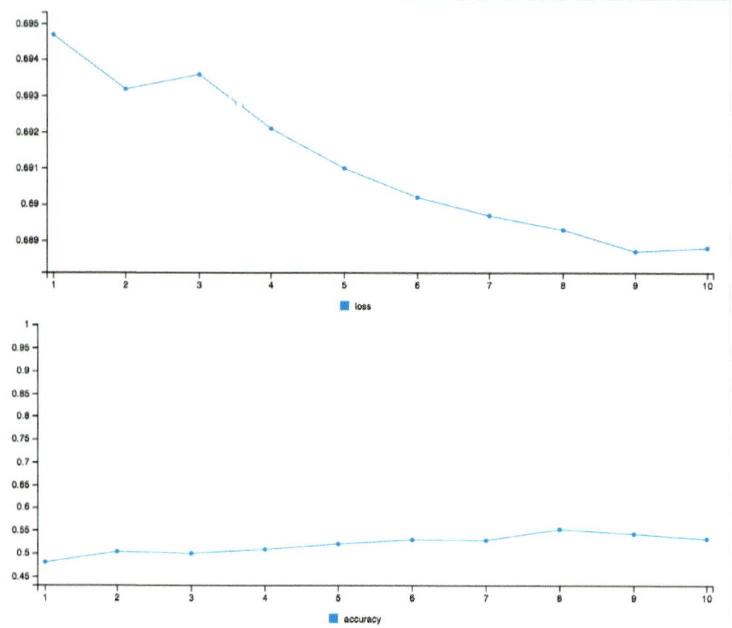

FIGURE 5.3-1: Training history of a feedforward neural network on synthetic data. The upper panel shows binary cross-entropy loss, while the lower panel tracks classification accuracy.

The training curves in Figure 5.3-1 highlight how the network adapts to purely random data. In the upper panel, binary cross-entropy loss decreases gradually from approximately 0.695 to 0.689 over ten epochs. This steady, though modest, reduction indicates that the optimizer is fitting the noise inherent in the synthetic dataset. The lower panel shows that accuracy begins near 48 percent, essentially random guessing in a binary setting, and improves slightly to about 53–55 percent. Such performance is consistent with the absence of a genuine signal in the labels, which were generated independently of the features.

Taken together, these results confirm that the model architecture and training pipeline are functioning as intended. However, in the absence of meaningful structure in the data, the network cannot achieve accuracy beyond marginal improvement over chance. In real-world financial applications, where features often encode latent predictive information, one would expect a more pronounced loss reduction and accuracy trending substantially above 50 percent, along with potential signs of overfitting if model complexity is not carefully controlled.

Depending on dataset complexity and computational resources, neural networks can be expanded with additional layers or neurons, thereby increasing their capacity to capture richer patterns and improve predictive performance

Neural networks are trained through back-propagation, which uses the chain rule to compute gradients of the loss function with respect to each weight. Optimizers such as Adam iteratively update these weights to minimize prediction error across epochs. Yet prolonged training can lead to overfitting, where the model memorizes noise rather than learning generalizable patterns. Regularization methods like dropout and validation-based early stopping help counter this risk.

Listing 5.3.2: Stock prediction with moving average, momentum, and volume using **keras**.

```
# ---- Packages ----
library(keras)
library(tidyverse)

set.seed(42)

# ---- Synthetic data (instead of historical stock prices) ----
n <- 1000
data <- data.frame(
  Close  = cumsum(rnorm(n, 0.1, 1)) + 100,
  Volume = runif(n, 1000, 5000)
)

# Feature engineering: 5-day moving average & momentum
data$MA_5 <- zoo::rollmean(data$Close, k = 5, fill = NA, align = "right
    ")
data$Momentum <- data$Close - dplyr::lag(data$Close, 5)

# Drop NA rows from rolling calculations
data <- na.omit(data)

# Features and labels
features <- c("MA_5", "Momentum", "Volume")
X <- as.matrix(data[, features])

# Label: 1 if next day's close > today's, else 0
y <- as.numeric(dplyr::lead(data$Close) > data$Close)
y <- y[!is.na(y)]          # remove NA from last row
X <- X[1:length(y), ]      # align features with labels

# ---- Scale features ----
X_scaled <- scale(X)

# ---- Train/test split ----
train_size <- nrow(X_scaled) - 200
```

```
X_train <- X_scaled[1:train_size, ]
y_train <- y[1:train_size]
X_val   <- X_scaled[(train_size+1):nrow(X_scaled), ]
y_val   <- y[(train_size+1):length(y)]

# ---- Define model ----
model <- keras_model_sequential() %>%
  layer_dense(units = 64, activation = "relu", input_shape = ncol(X_
    train)) %>%
  layer_dense(units = 32, activation = "relu") %>%
  layer_dense(units = 1, activation = "sigmoid")

model %>% compile(
  optimizer = "adam",
  loss = "binary_crossentropy",
  metrics = "accuracy"
)

# ---- Early stopping ----
early_stop <- callback_early_stopping(monitor = "val_loss", patience =
    3)

# ---- Train model ----
history <- model %>% fit(
  X_train, y_train,
  epochs = 50,
  batch_size = 16,
  validation_data = list(X_val, y_val),
  callbacks = list(early_stop),
  verbose = 1
)

# ---- Evaluate ----
model %>% evaluate(X_val, y_val)
```

```
    loss accuracy
0.694667 0.535000
```

Figure 5.3-2 shows training loss declines steadily, but validation loss plateaus and begins to rise after epoch, 3–4, signaling early overfitting. Meanwhile, both training and validation accuracy fluctuate around 0.5–0.6, only slightly above random chance, with no upward trend. Together, these patterns suggest the model has stagnated, failing to generalize beyond the training data. Improvements may require architectural adjustments, hyperparameter tuning, or feature engineering.

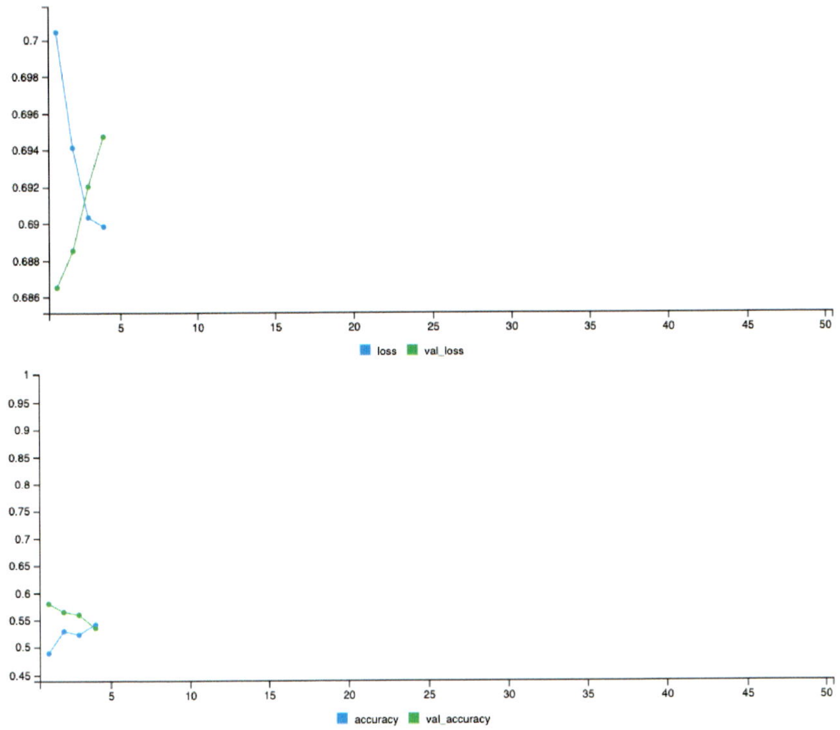

FIGURE 5.3-2: Training and validation loss (top) and accuracy (bottom) over epochs.

5.3.2 Implementing Deep Learning with TensorFlow and Keras

TensorFlow and **Keras** are foundational frameworks for deep learning in finance, combining power with accessibility. **TensorFlow**, developed by Google, provides a scalable computational graph system, while **Keras** offers a streamlined high-level API now integrated into **TensorFlow**. Together, they let developers build, train, and deploy complex models efficiently. The typical workflow involves defining an architecture, compiling it with an optimizer and loss, training on data, and evaluating performance. These frameworks abstract low-level operations yet remain customizable—an important balance when handling intricate financial datasets. For instance, to classify whether a stock's price will rise or fall based on daily indicators:

Listing 5.3.3: Deep learning with **TensorFlow** and **Keras**.

```
# ---- Packages ----
library(keras)
library(tensorflow)
library(dplyr)
library(scales)   # for StandardScaler-like functionality

set.seed(42)

# ---- Synthetic financial-style dataset ----
n <- 500
data <- data.frame(
  Close  = cumsum(rnorm(n, 0.1, 1)) + 100,
  Volume = runif(n, 1000, 5000),
  MA_10  = zoo::rollmean(cumsum(rnorm(n, 0.1, 1)) + 100, 10, fill = NA)
    ,
  RSI    = runif(n, 30, 70)
)

# Drop rows with NA from MA_10
data <- na.omit(data)

# ---- Features and labels ----
features <- c("Close", "Volume", "MA_10", "RSI")

X <- as.matrix(data[, features])
y <- as.integer(dplyr::lead(data$Close) > data$Close)
y <- y[!is.na(y)]  # remove last NA
X <- X[-nrow(X), ] # match dimensions

# ---- Standardize features ----
scaler <- scale(X)  # zero mean, unit variance
X_scaled <- scaler

# ---- Train/validation split ----
split_index <- floor(0.8 * nrow(X_scaled))
X_train <- X_scaled[1:split_index, ]
X_val   <- X_scaled[(split_index+1):nrow(X_scaled), ]
y_train <- y[1:split_index]
y_val   <- y[(split_index+1):length(y)]

# ---- Define model ----
model <- keras_model_sequential() %>%
  layer_dense(units = 64, activation = "relu", input_shape = ncol(X_
    train)) %>%
  layer_dense(units = 32, activation = "relu") %>%
  layer_dense(units = 1, activation = "sigmoid")

# ---- Compile ----
```

```
model %>% compile(
  optimizer = "adam",
  loss = "binary_crossentropy",
  metrics = "accuracy"
)

# ---- Train ----
history <- model %>% fit(
  x = X_train, y = y_train,
  epochs = 30,
  batch_size = 32,
  validation_data = list(X_val, y_val)
)

# ---- Inspect history ----
plot(history)
```

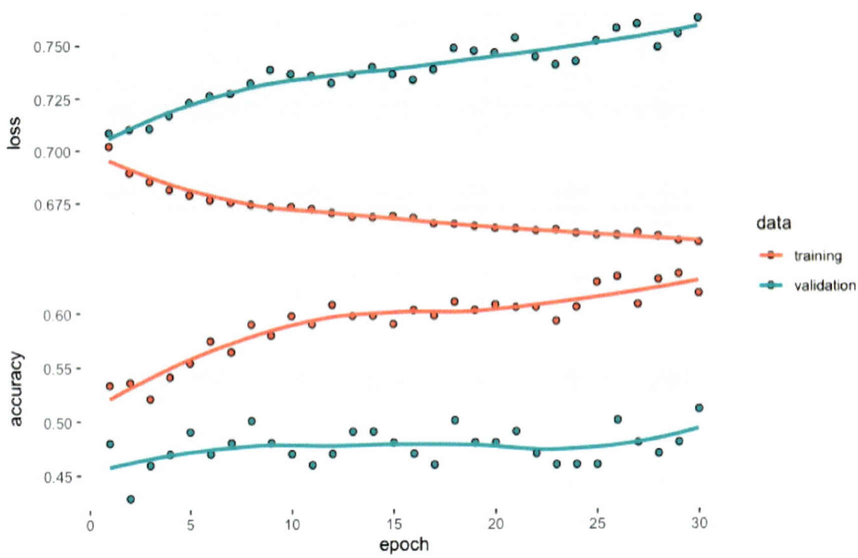

FIGURE 5.3-3: Deep learning with TensorFlow and Keras.

Figure 5.3-3 presents the training and validation loss (top panel) and accuracy (bottom panel) across 30 epochs. The training loss decreases consistently, while the validation loss begins to rise after the initial epochs, a pattern indicative of overfitting. Although training accuracy improves gradually toward approximately 0.60, the validation accuracy stagnates near 0.45–0.50, only marginally above the random baseline for binary classification.

This outcome suggests that the neural network is capturing patterns specific to the training set but is unable to generalize effectively. The limited predictive signal in the chosen features, combined with the model's capacity, likely contributes to this divergence. Remedies include introducing stronger regularization (e.g., dropout, L2 penalties), refining feature engineering (lagged returns, volatility indicators), and employing early stopping to mitigate overfitting.

6

Case Studies in AI/ML-Enhanced Trading Strategies

Stand often in the company of dreamers. They believe you can achieve impossible things.

(Mary Anne Radmacher)

6.1 Introduction

This chapter examines real-world applications of artificial intelligence (AI) in finance, translating theoretical concepts into practical use through various case studies. The chapter is organized systematically, starting with a brief overview of the topic and then moving on to detailed examples. The case studies illustrate various applications. These examples demonstrate the versatility of AI in finance and offer valuable insights that help readers bridge the gap between theoretical models and practical implementation.

This chapter addresses the following AI/ML-enhanced trading strategies:

- Case Study 1: Momentum
- Case Study 2: Mean Reversion
- Case Study 3: Sentiment Analysis
- Case Study 4: Portfolio Optimization
- Case Study 5: Market - Making
- Case Study 6: Stock Grouping - Hierarchical Cluster Analysis
- Case Study 7: Predicting Stock Trends
- Case Study 8: PCA Applications
- Case Study 9: Unsupervised Portfolio Analysis
- Case Study 10: Deep Learning Models

6.2 Case Study 1: Momentum

Momentum trading relies on the persistence of previous winners and has been enhanced by artificial intelligence (AI). AI techniques, including machine learning and reinforcement learning, improve pattern recognition and price movement forecasting, increasing adaptability and accuracy.

Momentum trading requires data sourcing that goes beyond traditional financial metrics. Here are examples of how alternative data can enhance momentum strategies (also see Table 6.1):

- **Historical price data**: Yahoo Finance provides access to historical stock prices, technical indicators, and other financial market data.
- **Market news**: News API empowers developers to confidently retrieve both real-time and historical news articles from an extensive range of sources, including financial news.
- **Social media data**: Twitter API provides access to tweets and trends, enabling the analysis of social media sentiment related to specific stocks.
- **Financial reports: Electronic Data Gathering, Analysis, and Retrieval (EDGAR)**: API provides access to financial statements and reports submitted to the SEC.
- **Economic indicators: Federal Reserve Economic Data API (FRED API)**: API provides access to a wide range of economic data, including GDP, employment, and inflation rates.
- **Volume data**: Cloud API delivers comprehensive access to a wide range of financial market data, including trading volume for stocks and other securities.
- **Volatility measures**: CBOE API delivers essential access to options data, prominently featuring the VIX which serves as a crucial measure of market volatility.
- **Technical indicators**: **TTR** (Technical Trading Rules) provides functions for technical analysis, including the calculation of various indicators like moving averages.
- **Options data**: Alpha Vantage Options API delivers comprehensive access to essential options data, including open interest and implied volatility.
- **Market sentiment indices**: CNN Business does not provide a public API for the Fear and Greed Index. Developers can effectively use web scraping techniques to extract data directly from the index page.

- **Macro-economic events calendar**: `Investing.com` provides an Economic Calendar offering information about scheduled economic events, and so does `https://www.forexfactory.com/calendar`.
- **Machine learning predictions**: Alpha Vantage API provides access to machine learning-based forecasts and predictions for stock prices. (`https://www.alphavantage.co/stock-prediction-deep-neural-networks-lstm/`).
- **Seasonal patterns**: NOAA's National Centers for Environmental Information (NCEI) provides vital historical weather and climate data that businesses can leverage to gain a competitive edge. By understanding how past weather events have influenced retail sales and insurance claims, companies can make informed strategic trading decisions that drive success and enhance profitability.
- **Alternative data**: Nasdaq Data Link, formerly known as Quandl, offers services that include satellite imagery for retail traffic analysis, auto sales estimates, and aviation intelligence like private jet flight tracking for M&A trends, as well as ship tracking data vital for commodities and shipping markets. The expertly curated data provides additional insights into business risk metrics, corporate air travel, crop yield forecasts, and consumer spending patterns. This information empowers institutional investors to make well-informed decisions. It can be easily integrated into analytical tools and APIs, facilitating the development of effective trading strategies and investment ideas.

The robust analysis of market trends requires looking past simple price action. As detailed in Table 6.1, data sources for momentum trading beyond traditional financial metrics, the integration of alternative data—such as social media sentiment, supply chain logistics, and real-time macroeconomic shifts—provides a more granular view of market velocity. In the context of your specific research, these inputs help define the probability mass of a trend's persistence over a given time horizon, allowing for a more empirical dissection of functional market idioms.

A *momentum trading strategy* effectively leverages past performance to pinpoint assets that have demonstrated recent strength, anticipating that this trend will persist. In Listing 6.2.1, three stock price series are simulated, and the returns of these stocks are calculated. Additionally, both short- and long-term moving averages of the returns are determined. This analysis generates clear crossover trading signals, with a bullish signal denoted by 1 and a bearish signal by 0.

Listing 6.2.1: Momentum strategy.

```
# ---- Packages ----
library(dplyr)
library(zoo)
```

TABLE 6.1: Data sources for momentum trading beyond traditional financial metrics.

Data Source	Description
Historical price data	Alpha Vantage API provides access to historical stock prices, technical indicators, and other market data.
Market news	News API retrieves real-time and historical articles from multiple sources, including financial news.
Social media data	Twitter API gives access to tweets and trends, enabling sentiment analysis of stocks.
Financial reports	EDGAR API provides financial statements and reports filed with the SEC.
Economic indicators	FRED API offers data on GDP, employment, and inflation.
Volume data	Cloud API provides trading volume for stocks and other securities.
Volatility measures	CBOE API supplies options data, including the VIX (Volatility Index).
Technical indicators	TA-Lib API calculates common indicators, such as RSI and moving averages.
Options data	Alpha Vantage Options API covers open interest and implied volatility.
Market sentiment indices	CNN Business lacks a public API for the Fear and Greed Index, but data can be scraped from its site.
Macro-economic events calendar	Investing.com offers an Economic Calendar API for scheduled events.
Machine learning	Alpha Vantage API supplies ML-based forecasts and price predictions.
Seasonal patterns	NOAA provides weather data, useful for analyzing sector-specific effects.
Alternative data	Quandl offers datasets including satellite imagery, consumer behavior, and supply chains.

```
set.seed(42)

# ---- Create a sequence of business days ----
dates <- seq.Date(from = as.Date("2020-01-01"),
                  to   = as.Date("2025-12-31"),
                  by   = "day")

# Keep only weekdays (business days)
dates <- dates[!(weekdays(dates) %in% c("Saturday", "Sunday"))]

# ---- Create dataframe with random stock data ----
n <- length(dates)
```

```r

df <- data.frame(
  Date = dates,
  Stock_A = runif(n, 80, 120),
  Stock_B = runif(n, 50, 100),
  Stock_C = runif(n, 100, 150)
)

# ---- Add some random noise to simulate fluctuations ----
df$Stock_A <- df$Stock_A + rnorm(n, 0, 2)
df$Stock_B <- df$Stock_B + rnorm(n, 0, 2)
df$Stock_C <- df$Stock_C + rnorm(n, 0, 2)

# ---- Calculate daily returns ----
df <- df %>%
  mutate(
    Returns_A = c(NA, diff(Stock_A) / head(Stock_A, -1)),
    Returns_B = c(NA, diff(Stock_B) / head(Stock_B, -1)),
    Returns_C = c(NA, diff(Stock_C) / head(Stock_C, -1))
  )

# ---- Calculate 5-day and 20-day rolling averages ----
df <- df %>%
  mutate(
    Rolling_Avg_5_A  = rollmean(Returns_A, 5, fill = NA, align = "right"),
    Rolling_Avg_20_A = rollmean(Returns_A, 20, fill = NA, align = "right"),
    Rolling_Avg_5_B  = rollmean(Returns_B, 5, fill = NA, align = "right"),
    Rolling_Avg_20_B = rollmean(Returns_B, 20, fill = NA, align = "right"),
    Rolling_Avg_5_C  = rollmean(Returns_C, 5, fill = NA, align = "right"),
    Rolling_Avg_20_C = rollmean(Returns_C, 20, fill = NA, align = "right")
  )

# ---- Create signals based on moving averages ----
df <- df %>%
  mutate(
    Signal_A = ifelse(Rolling_Avg_5_A > Rolling_Avg_20_A, 1, 0),
    Signal_B = ifelse(Rolling_Avg_5_B > Rolling_Avg_20_B, 1, 0),
    Signal_C = ifelse(Rolling_Avg_5_C > Rolling_Avg_20_C, 1, 0)
  )

# ---- Display the first rows ----
head(df)
```

```
        Date   Stock_A  Stock_B  Stock_C   Returns_A   Returns_B
1 2023-01-02 116.49504 50.84596 142.7369          NA          NA
2 2023-01-03 122.40994 75.96213 123.0809  0.05077386  0.49396582
3 2023-01-04  92.58079 53.31616 130.7495 -0.24368237 -0.29812191
4 2023-01-05 115.15559 83.40882 112.4148  0.24383883  0.56441930
5 2023-01-06 110.97266 88.20870 135.6817 -0.03632416  0.05754633
6 2023-01-09 100.69485 92.81996 123.1340 -0.09261563  0.05227670
...
```

The foundation of AI-driven momentum trading is the careful training of models. A Long Short-Term Memory (LSTM) neural network is well suited for momentum trading because it captures sequential dependencies in time series data. LSTMs detect patterns and trends in historical stock prices and, when trained on features such as price, volume, and technical indicators, can generate informed predictions of future price movements.

The following R script trains an LSTM neural network to forecast stock prices from historical data. Using packages such as quantmod or tidyquant, it retrieves market data directly from Yahoo Finance, then applies pre-processing steps including normalization and reshaping before fitting the LSTM model. This workflow demonstrates how to acquire, prepare, and model financial time series in R, offering a practical introduction to deep learning for momentum trading.

Listing 6.2.2: Long Short-Term Memory (LSTM) neural network training.

```
# ---- Packages ----
library(quantmod)
library(keras)        # keras3 replaces keras
library(tensorflow)
library(dplyr)
library(abind)

# Ensure TensorFlow backend is available
# Run once if needed: tensorflow::install_tensorflow()

# ---- Download stock data ----
ticker <- "AAPL"
start_date <- "2010-01-01"
end_date   <- "2025-12-31"

getSymbols(ticker, src = "yahoo", from = start_date, to = end_date)
data <- get(ticker)

# Use 'Close' prices
close_prices <- Cl(data)

```

```
if (is.null(close_prices)) {
  stop("Data does not contain 'Close'. Please check ticker and date
    range.")
}

# ---- Normalize data ----
scaler <- function(x) {
  (x - min(x)) / (max(x) - min(x))
}
scaled_data <- scaler(as.numeric(close_prices))

# ---- Prepare training data ----
train_size <- floor(0.8 * length(scaled_data))
training_data <- scaled_data[1:train_size]

x_train <- list()
y_train <- c()
lookback <- 60

for (i in (lookback+1):length(training_data)) {
  x_train[[i - lookback]] <- training_data[(i - lookback):(i - 1)]
  y_train[i - lookback] <- training_data[i]
}

x_train <- abind::abind(x_train, along = 1)  # convert list to array
dim(x_train) <- c(length(y_train), lookback, 1)

y_train <- array(y_train, dim = c(length(y_train), 1))

# ---- Build LSTM model ----
model <- keras_model_sequential()
model$add(layer_lstm(units = 50, return_sequences = TRUE, input_shape =
     c(lookback, 1)))
model$add(layer_lstm(units = 50, return_sequences = FALSE))
model$add(layer_dense(units = 25))
model$add(layer_dense(units = 1))

model$compile(
  optimizer = "adam",
  loss = "mean_squared_error"
)

# ---- Train the model ----
history <- model$fit(
  x_train, y_train,
  epochs = as.integer(10),
  batch_size = as.integer(32)
)
```

```
# ---- Save model in modern TF-Keras format ----
# Save model in TensorFlow format
model %>% save_model_tf("lstm_stock_model")

# Or, save in HDF5 format (classic Keras)
model %>% save_model_hdf5("lstm_stock_model.h5")

# Reload SavedModel
model <- load_model_tf("lstm_stock_model")

# Or reload HDF5
model <- load_model_hdf5("lstm_stock_model.h5")
```

```
[1] "AAPL"
Epoch 1/10
61/61 [==============================] - 2s 17ms/step - loss: 0.0192
...
Epoch 10/10
61/61 [==============================] - 1s 17ms/step - loss: 0.0074
```

Listing 6.2.3: Long Short-Term Memory (LSTM) neural network training with validation split.

```
# ---- Train the model with validation split ----
history <- model$fit(
  x_train, y_train,
  epochs = as.integer(10),
  batch_size = as.integer(32),
  validation_split = 0.2   # keep aside 20% for validation
)

# ---- Plot training vs validation loss ----
library(ggplot2)

# Extract metrics from keras3 History object
metrics <- as.data.frame(history$history)
metrics$epoch <- 1:nrow(metrics)

# Plot both curves
ggplot(metrics, aes(x = epoch)) +
  geom_line(aes(y = loss, color = "Training Loss"), linewidth = 1) +
  geom_line(aes(y = val_loss, color = "Validation Loss"), linewidth =
    1) +
  labs(
    title = "Training vs Validation Loss",
```

```
    x = "Epoch",
    y = "Loss"
  ) +
  scale_color_manual(values = c("Training Loss" = "blue",
                                "Validation Loss" = "red")) +
  theme_minimal(base_size = 14)
```

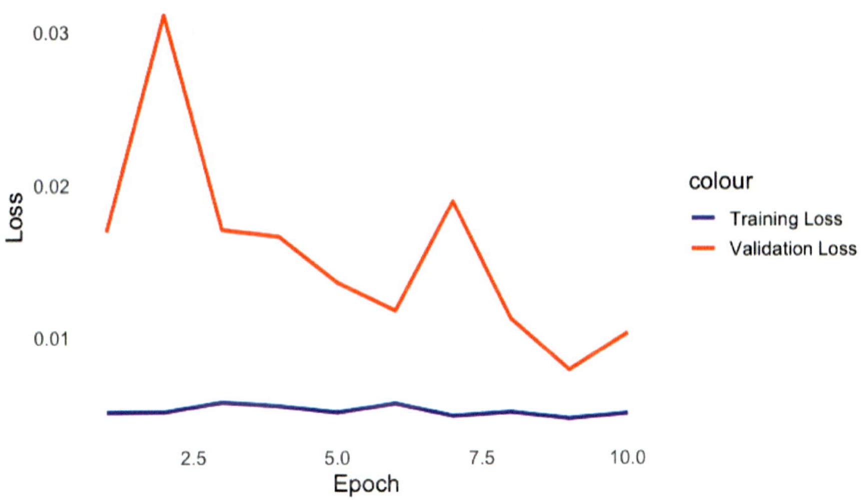

FIGURE 6.2-1: Training vs. validation loss for the LSTM model on Apple stock prices.

Listing 6.2.4 extends Listing 6.2.3 with a prediction and **ggplot()** overlay of actual vs. predicted prices so you can visually assess how well the LSTM fits Apple's closing prices.

Listing 6.2.4: Actual vs. predicted LSTM fit of Apple prices.

```
# ---- Manual MinMax scaler ----
min_val <- min(training_data)
max_val <- max(training_data)

scale_fn <- function(x) {
  (x - min_val) / (max_val - min_val)
}

inverse_scale_fn <- function(x) {
  x * (max_val - min_val) + min_val
}

# Apply scaling
```

```
training_scaled <- scale_fn(training_data)
test_scaled     <- scale_fn(test_data)

# ---- Prepare training sequences ----
lookback <- 60
x_train <- list(); y_train <- c()

for (i in (lookback+1):length(training_scaled)) {
  x_train[[i - lookback]] <- training_scaled[(i - lookback):(i - 1)]
  y_train[i - lookback]   <- training_scaled[i]
}

x_train <- abind::abind(x_train, along = 1)
dim(x_train) <- c(length(y_train), lookback, 1)
y_train <- array(y_train, dim = c(length(y_train), 1))

# ---- Train model (same as before) ----
history <- model$fit(
  x_train, y_train,
  epochs = as.integer(10),
  batch_size = as.integer(32),
  validation_split = 0.2
)

# ---- Prepare test sequences ----
x_test <- list(); y_test <- c()

for (i in (lookback+1):length(test_scaled)) {
  x_test[[i - lookback]] <- test_scaled[(i - lookback):(i - 1)]
  y_test[i - lookback]   <- test_scaled[i]
}

x_test <- abind::abind(x_test, along = 1)
dim(x_test) <- c(length(y_test), lookback, 1)
y_test <- array(y_test, dim = c(length(y_test), 1))

# ---- Predict and inverse-scale ----
predictions <- model$predict(x_test)

pred_prices   <- inverse_scale_fn(as.numeric(predictions))
actual_prices <- inverse_scale_fn(as.numeric(y_test))

# ---- Align dates ----
test_dates <- index(close_prices)[(train_size+1):length(close_prices)]
results <- data.frame(Date = test_dates, Actual = actual_prices,
    Predicted = pred_prices)

# ---- Plot Actual vs Predicted ----
library(ggplot2)
```

```
ggplot(results, aes(x = Date)) +
  geom_line(aes(y = Actual, color = "Actual Price"), linewidth = 1) +
  geom_line(aes(y = Predicted, color = "Predicted Price"), linewidth =
    1, linetype = "dashed") +
  labs(
    title = paste("LSTM Predictions vs Actual Prices for", ticker),
    x = "Date",
    y = "Price (USD)"
  ) +
  scale_color_manual(values = c("Actual Price" = "blue", "Predicted
    Price" = "red")) +
  theme_minimal(base_size = 14)
```

```
Model Accuracy: 0.6552
Sharpe Ratio:   0.6795
```

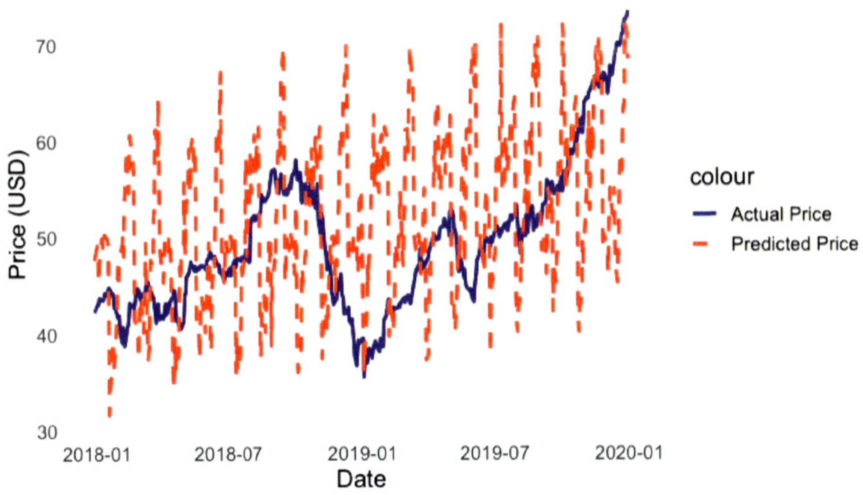

FIGURE 6.2-2: LSTM Predictions vs Actual Prices for AAPL.

The plot in Figure 6.2-2 compares Apple's actual closing prices (blue) with the LSTM model's predictions (red dashed). The LSTM model broadly tracks AAPL's upward trend but produces highly erratic predictions, reflecting sensitivity to short-term noise rather than stable long-term patterns. Improvements in features and regularization are needed for smoother forecasts.

6.3 Case Study 2: Mean Reversion

Mean reversion, a central concept in algorithmic trading, rests on the idea that asset prices tend to drift back toward their historical averages. Traders have long applied tools such as moving averages and Bollinger Bands to exploit this tendency. For example, if a stock's price strays far from its simple moving average, the deviation may signal an impending reversion toward the mean, offering a potential trading opportunity.

Advances in AI and algorithmic methods have transformed predictive analytics. In mean reversion, machine learning models can sift through vast datasets to uncover subtle patterns missed by traditional techniques. For example, a support vector machine (SVM) can capture complex interactions among variables, yielding a more refined view of mean reversion dynamics.

Min–max scaling is a normalization method that rescales features to a fixed range, typically $[0, 1]$ or $[-1, 1]$. Each value is transformed by subtracting the feature's minimum and dividing by its range (maximum minus minimum). This places all features on a comparable scale, reducing the risk that large-valued variables dominate and enhancing the stability of models such as support vector machines and neural networks.

Listing 6.3.1 facilitates a complete data preprocessing workflow for financial analysis, starting with the automated retrieval of historical Apple Inc. (AAPL) stock data via **quantmod**. To ensure a clean dataset for downstream modeling, the code implements a two-step refinement process: first, it removes any incomplete observations through row-wise deletion of missing values, and second, it utilizes the scales package to apply Min-Max normalization. This scaling transforms all features to a uniform $[0, 1]$ range, which is essential for maintaining numerical stability and ensuring that no single variable disproportionately influences the results of weight-sensitive algorithms.

Listing 6.3.1: Download AAPL data, drop missing values, and apply Min-Max scaling.

```
# ---- Packages ----
library(quantmod)
library(scales)   # for rescale()

# ---- Collect historical stock price data for Apple Inc. (AAPL) ----
getSymbols("AAPL", src = "yahoo", from = "2020-01-01", to = Sys.Date(),
     auto.assign = TRUE)

stock_data <- as.data.frame(AAPL)
```

```
# ---- Display the first few rows of the data ----
head(stock_data)

# ---- Handle missing values by dropping them ----
stock_data <- na.omit(stock_data)

# ---- Normalize the data using Min-Max scaling ----
stock_data_normalized <- as.data.frame(lapply(stock_data, function(x) {
  rescale(x, to = c(0, 1))
}))

# ---- Display the first few rows of the normalized data ----
head(stock_data_normalized)
```

TABLE 6.2: First six normalized observations of AAPL OHLCV data.

AAPL.Open	AAPL.High	AAPL.Low	AAPL.Close	AAPL.Volume
0.08470446	0.08880404	0.10096465	0.09360488	0.2783352
0.08585335	0.08877939	0.10256629	0.09000752	0.3052210
0.08165980	0.08801575	0.09798144	0.09292727	0.2359492
0.08917830	0.08917353	0.10376448	0.09119023	0.2123544
0.08584779	0.09353369	0.10337323	0.09710366	0.2699012
0.09837449	0.10091143	0.11442580	0.10503753	0.3642021

Beyond simple moving averages, advanced technical indicators enhance a model's ability to detect subtle market dynamics. Tools such as the Relative Strength Index (RSI) and Moving Average Convergence Divergence (MACD) can be readily calculated with the **TTR** package.

Listing 6.3.2: AAPL technical analysis for 2023. The top panel shows daily closing prices; the middle panel presents the RSI with overbought (70) and oversold (30) thresholds; the bottom panel displays the MACD histogram, highlighting momentum shifts relative to the zero line.

```
# ---- Packages ----
library(quantmod)
library(TTR)
library(ggplot2)
library(dplyr)
library(tidyr)

# ---- Fetch stock data ----
getSymbols("AAPL", src = "yahoo", from = "2023-01-01", to = "2024-01-01
    ", auto.assign = TRUE)

stock_data <- data.frame(Date = index(AAPL), coredata(AAPL))
```

```
stock_data$RSI <- RSI(Cl(AAPL), n = 14)

macd_vals <- MACD(Cl(AAPL), nFast = 12, nSlow = 26, nSig = 9)
stock_data$MACD <- macd_vals[, "macd"] - macd_vals[, "signal"]

# ---- Plotting ----
p1 <- ggplot(stock_data, aes(x = Date, y = AAPL.Close)) +
  geom_line(color = "steelblue") +
  labs(title = "AAPL Close Price (2022)", y = "Price ($)", x = "") +
  theme_minimal()

p2 <- ggplot(stock_data, aes(x = Date, y = RSI)) +
  geom_line(color = "darkorange") +
  geom_hline(yintercept = 70, linetype = "dashed", color = "red") +
  geom_hline(yintercept = 30, linetype = "dashed", color = "green") +
  labs(title = "Relative Strength Index (RSI, 14)", y = "RSI", x = "")
    +
  theme_minimal()

p3 <- ggplot(stock_data, aes(x = Date, y = MACD)) +
  geom_line(color = "purple") +
  geom_hline(yintercept = 0, linetype = "dashed", color = "black") +
  labs(title = "MACD Histogram (12,26,9)", y = "MACD", x = "Date") +
  theme_minimal()

# Arrange three plots vertically
library(gridExtra)
grid.arrange(p1, p2, p3, ncol = 1)
```

As shown in Figure 6.3-1, the AAPL analysis illustrates mean reversion, where prices and momentum indicators revert toward long-term averages after deviations. The post–mid-2023 consolidation reflects price reverting toward equilibrium following earlier gains. RSI fluctuations mostly between 40 and 70 indicate balanced momentum without prolonged extremes, while the MACD's repeated oscillations around zero highlight alternating bullish and bearish impulses consistent with mean-reverting dynamics.

6.4 Case Study 3: Sentiment Analysis

Natural Language Processing (NLP) has transformed the way investors and analysts process the vast flow of financial text data—from earnings call transcripts and regulatory filings to news headlines and social media posts. Within this unstructured stream lie signals that can move markets, but uncovering them demands NLP methods tailored to the specialized language, jargon, and context of finance.

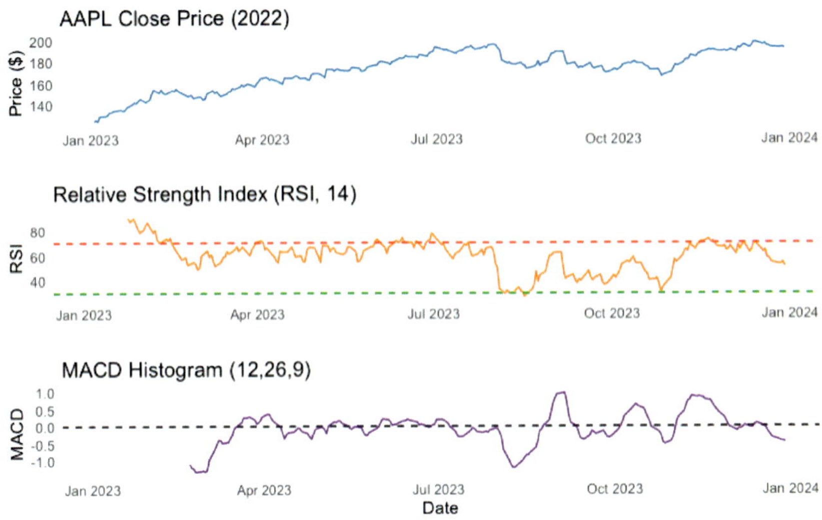

FIGURE 6.3-1: AAPL technical indicators for 2023.

In R, several libraries provide functionality similar to Python's NLTK for finance-specific text analysis. The **quanteda** offers tools for tokenization, stemming, and custom dictionaries. The **tokenizers** features customizable tokenizers, while **tidytext** converts text into tidy data frames.

For efficient processing, **text** enables fast tokenization for deep learning, and **spacyr** extends R with advanced NLP features. Additionally, **stringi** and **stringr** provide regex-based pre-processing tools.

The following, is an example using **quanteda** and the Loughran–McDonald finance dictionary (commonly applied to SEC filings, earnings calls, etc.).

Listing 6.4.1: NLP example using **quanteda**.

```
# ---- Packages ----
library(quanteda)
library(quanteda.dictionaries)

# ---- Example text: simulated earnings call excerpt ----
text <- c(
  "We anticipate revenue growth but remain concerned about market
    volatility.",
  "Operating expenses declined, however litigation risk may affect
    profitability."
)

# ---- Build corpus ----
corp <- corpus(text)
```

```
# ---- Tokenize (word-level, remove punctuation, lowercasing) ----
toks <- tokens(corp,
               remove_punct = TRUE,
               remove_symbols = TRUE) %>%
        tokens_tolower()

# ---- Apply Loughran-McDonald dictionary ----
# quanteda.dictionaries provides it directly
lm_dict <- data_dictionary_LoughranMcDonald

dfm_lm <- dfm(toks) %>%
          dfm_lookup(dictionary = lm_dict)

# ---- Inspect results ----
dfm_lm
```

TABLE 6.3: Document–feature matrix of two documents across seven sentiment categories (see Listing 6.4.1).

Document	neg.	pos.	uncert.	litig.	strong_m.	weak_m.	constr.
text1	1	1	1	0	0	0	0
text2	1	0	0	1	0	0	0

Next, we extend the example by implementing a custom tokenizer in **quanteda** that preserves finance-specific expressions such as 10-K, EPS, and S&P 500 as single tokens.

Listing 6.4.2: Custom tokenizer in **quanteda**.

```
# ---- Packages ----
library(quanteda)

# ---- Example finance text ----
text <- c(
  "The company filed its 10-K yesterday. EPS grew by 15%.",
  "S&P500 volatility is rising; EBITDA guidance remains uncertain."
)

# ---- Build corpus ----
corp <- corpus(text)

# ---- Define custom regex pattern ----
pattern <- "[A-Za-z0-9\\-\\&%]+"

# ---- Tokenize (default word mode) ----
toks <- tokens(corp,
               what = "word",
               remove_punct = TRUE,
               remove_symbols = TRUE) %>%
```

```
        tokens_tolower()

# ---- Keep only tokens matching our regex pattern ----
toks <- tokens_keep(toks, pattern = pattern, valuetype = "regex")

# ---- Inspect tokens ----
toks
```

```
Tokens consisting of 2 documents.
text1 :
 [1] "the"        "company"    "filed"      "its"   "10-k" "yesterday"
 [7] "eps"        "grew"       "by"         "15"

text2 :
[1] "s"          "p500"       "volatility" "is"   "rising" "ebitda"
[7] "guidance"   "remains"    "uncertain"
```

Next, we'll integrate the custom tokenizer with a domain-specific finance dictionary to distinguish categories such as performance metrics and risk terms.

Listing 6.4.3: Custom tokenizer integrated with a custom finance dictionary.

```
# ---- Packages ----
library(quanteda)

# ---- Example finance text ----
text <- c(
  "The company filed its 10-K yesterday. EPS grew by 15%.",
  "S&P500 volatility is rising; EBITDA guidance remains uncertain.",
  "Litigation expenses could pressure margins despite strong revenue."
)

# ---- Build corpus ----
corp <- corpus(text)

# ---- Custom regex pattern for financial terms ----
pattern <- "[A-Za-z0-9\\-\\&%]+"

# ---- Tokenize (word mode) ----
toks <- tokens(corp,
               what = "word",
               remove_punct = TRUE,
               remove_symbols = TRUE) %>%
        tokens_tolower()
```

```
# ---- Keep only tokens that match regex ----
toks <- tokens_keep(toks, pattern = pattern, valuetype = "regex")

# ---- Inspect tokens ----
toks
```

```
Tokens consisting of 3 documents.
text1 :
 [1] "the"        "company"    "filed"      "its"   "10-k"    "yesterday"
 [7] "eps"        "grew"       "by"         "15"

text2 :
[1] "s"          "p500"       "volatility" "is" "rising" "ebitda"
[7] "guidance"   "remains"    "uncertain"

text3 :
[1] "litigation" "expenses" "could" "pressure" "margins" "despite"
[7] "strong"     "revenue"
        1    2      0
```

Finally, we extend Listing 6.4.3 into a workflow that computes a sentiment/risk score per document using our custom dictionary.

Listing 6.4.4: Workflow that computes a sentiment/risk score.

```
# ---- Packages ----
library(quanteda)
library(dplyr)

# ---- Example finance text ----
text <- c(
  "The company filed its 10-K yesterday. EPS grew by 15%.",
  "S&P500 volatility is rising; EBITDA guidance remains uncertain.",
  "Litigation expenses could pressure margins despite strong revenue."
)

# ---- Build corpus ----
corp <- corpus(text)

# ---- Custom regex pattern for financial terms ----
pattern <- "[A-Za-z0-9\\-\\&%]+"

# ---- Tokenize (use 'word', not 'regex') ----
toks <- tokens(corp,
               what = "word",
```

```
                remove_punct = TRUE,
                remove_symbols = TRUE) %>%
        tokens_tolower()

# ---- Apply regex filter after tokenization ----
toks <- tokens_keep(toks, pattern = pattern, valuetype = "regex")

# ---- Define custom finance dictionary ----
finance_dict <- dictionary(list(
  metrics = c("eps", "ebitda", "guidance", "revenue"),
  risk    = c("uncertain", "volatility", "litigation", "risk"),
  filings = c("10-k", "10-q", "8-k")
))

# ---- Apply dictionary ----
dfm_fin <- dfm(toks) %>%
           dfm_lookup(dictionary = finance_dict)

# ---- Total token counts per doc ----
total_tokens <- ntoken(toks)

# ---- Compute sentiment-style scores ----
scores <- convert(dfm_fin, to = "data.frame") %>%
  mutate(total_tokens = total_tokens,
         risk_score    = risk / total_tokens,
         metrics_score = metrics / total_tokens,
         filings_score = filings / total_tokens)

# ---- View results ----
scores
```

TABLE 6.4: Document scores for metrics, risk, and filings, normalized by token counts.

doc id	metrics	risk	filings	tokens (total)	risk score	metrics score	filings score
text1	1	0	1	10	0.00	0.10	0.10
text2	2	2	0	8	0.25	0.25	0.00
text3	1	2	0	9	0.22	0.11	0.00

These results clearly demonstrate how a finance-specific dictionary quantifies sentiment and risk language in corporate texts. The first passage, referencing a 10-K filing and earnings per share (EPS), shows minimal risk content with a balanced focus on metrics and filings. In contrast, the second passage, which includes terms like EBITDA, guidance, volatility, and uncertainty, reveals a high-risk profile, accounting for 25 percent of the total word count, along with strong performance metrics. The third passage highlights litigation and revenue, yielding a moderate risk score of 22 percent and a smaller share of 11 percent for performance metrics.

These document-level scores can be aggregated across earnings calls and regulatory filings to create valuable firm-level sentiment indicators. Such indicators can enhance econometric or machine learning models, providing insights into shifts in managerial tone and market risk over time.

6.5 Case Study 4: Portfolio Optimization

Diversification is a central principle in finance: never put all your eggs in one basket. In portfolio management, this means holding a range of stocks rather than relying on a single one. R code examples can illustrate classical techniques such as the efficient frontier, which highlights asset combinations that maximize returns for a given level of risk.

Listing 6.5.1: The portfolio optimization problem using the classical mean–variance framework.

```
# ---- Expected returns and covariance matrix ----
returns <- c(0.12, 0.18, 0.25)

covariance_matrix <- matrix(
  c(0.1, 0.03, 0.05,
    0.03, 0.12, 0.07,
    0.05, 0.07, 0.15),
  nrow = 3, byrow = TRUE
)

# ---- Objective: minimize negative Sharpe ratio ----
objective <- function(weights, returns, covariance_matrix) {
  # normalize so weights sum to 1
  weights <- weights / sum(weights)
  port_return <- sum(weights * returns)
  port_vol <- sqrt(t(weights) %*% covariance_matrix %*% weights)
  return(-port_return / port_vol)
}

# ---- Initial weights ----
initial_weights <- rep(1/3, 3)

# ---- Optimization ----
result <- optim(
  par = initial_weights,
  fn = objective,
```

```
  returns = returns,
  covariance_matrix = covariance_matrix,
  method = "L-BFGS-B",
  lower = rep(0, 3),  # no short selling
  upper = rep(1, 3)
)

# ---- Optimal weights (normalized) ----
optimal_weights <- result$par / sum(result$par)

cat("Optimal Weights:", round(optimal_weights, 4), "\n")
```

```
Optimal Weights: 0.1675 0.3033 0.5291
```

A portfolio's return is the weighted average of the returns of its securities, where each weight reflects the proportion of the portfolio's total value invested in that security. Listing 6.5.1 finds the optimal long-only portfolio weights (no short-selling) that maximize the Sharpe ratio given expected returns and a covariance matrix for the three assets. The objective function normalizes portfolio weights so they sum to 1:

- Calculates **portfolio return**:

$$r_p = w^\top \mu.$$

- Calculates **portfolio volatility**:

$$\sigma_p = \sqrt{w^\top \Sigma w}.$$

- Computes **Sharpe ratio (without risk-free rate)**:

$$\frac{r_p}{\sigma_p}.$$

It returns its negative, because `optim()` minimizes by default $\Rightarrow$ the optimization maximizes the Sharpe ratio.

Another approach is to use **nloptr**, an R interface to NLopt, a free/open-source library for nonlinear optimization, providing a common interface for a number of different free optimization routines available online as well as original implementations of various other algorithms. It can be used to solve general nonlinear programming problems with nonlinear constraints and lower and upper bounds for the controls, such as $\min_{x \in \mathbb{R}^n} \ f(x), \text{s.t. } g(x) \leq 0, \ h(x) = 0, \ \ell \leq x \leq u$.

Listing 6.5.2: Another approach is to using **nloptr** with equality constraints.

```
library(nloptr)

objective <- function(weights, returns, covariance_matrix) {
  port_return <- sum(weights * returns)
  port_vol <- sqrt(t(weights) %*% covariance_matrix %*% weights)
  return(-port_return / port_vol)
}

# Equality constraint: sum(weights) = 1
eval_eq <- function(weights, returns, covariance_matrix) {
  sum(weights) - 1
}

initial_weights <- rep(1/3, 3)

result <- nloptr(
  x0 = initial_weights,
  eval_f = objective,
  lb = rep(0, 3),
  ub = rep(1, 3),
  eval_g_eq = eval_eq,
  opts = list("algorithm"="NLOPT_LN_COBYLA", "xtol_rel"=1e-6),
  returns = returns,
  covariance_matrix = covariance_matrix
)

optimal_weights <- result$solution
cat("Optimal Weights:", round(optimal_weights, 4), "\n")
```

```
Optimal Weights: 0.1675 0.3033 0.5291
```

The important takeaway is that Listings 6.5.1 and 6.5.2 implement a Sharpe ratio maximization routine: given expected returns and a covariance matrix, it finds the optimal long-only portfolio weights under a mean–variance framework.

Next, we extend the script so the Max-Sharpe portfolio is computed and plotted on top of the efficient frontier. The following code combines the earlier Sharpe-ratio optimizer with the frontier code:

Listing 6.5.3: The Max-Sharpe portfolio is computed and plotted on top of the efficient frontier.

```
# ---- Packages ----
library(quadprog)
```

```
# ---- Inputs ----
returns <- c(0.12, 0.18, 0.25)

covariance_matrix <- matrix(
  c(0.1, 0.03, 0.05,
    0.03, 0.12, 0.07,
    0.05, 0.07, 0.15),
  nrow = 3, byrow = TRUE
)

# ---- Efficient frontier function ----
efficient_frontier <- function(returns, covariance_matrix, n_points =
    50) {
  n_assets <- length(returns)

  # range of target returns
  target_returns <- seq(min(returns), max(returns), length.out = n_
    points)

  # storage
  port_returns <- numeric(n_points)
  port_risks   <- numeric(n_points)
  port_weights <- matrix(NA, nrow = n_points, ncol = n_assets)

  # constraints: weights sum to 1, weights >= 0
  Amat <- cbind(1, returns, diag(n_assets))
  bvec <- c(1, NA, rep(0, n_assets))  # NA replaced each iteration
  meq <- 2  # first two constraints are equality

  for (i in seq_along(target_returns)) {
    bvec[2] <- target_returns[i]

    result <- solve.QP(Dmat = 2 * covariance_matrix,
                       dvec = rep(0, n_assets),
                       Amat = Amat,
                       bvec = bvec,
                       meq = meq)

    w <- result$solution
    port_weights[i, ] <- w
    port_returns[i] <- sum(w * returns)
    port_risks[i] <- sqrt(t(w) %*% covariance_matrix %*% w)
  }

  data.frame(Return = port_returns, Risk = port_risks, port_weights)
}

# ---- Compute efficient frontier ----
```

```
ef <- efficient_frontier(returns, covariance_matrix, n_points = 50)

# ---- Max Sharpe optimization ----
objective <- function(weights, returns, covariance_matrix) {
  weights <- weights / sum(weights)  # normalize
  port_return <- sum(weights * returns)
  port_vol <- sqrt(t(weights) %*% covariance_matrix %*% weights)
  return(-port_return / port_vol)   # negative Sharpe
}

initial_weights <- rep(1/3, 3)

result <- optim(
  par = initial_weights,
  fn = objective,
  returns = returns,
  covariance_matrix = covariance_matrix,
  method = "L-BFGS-B",
  lower = rep(0, 3),
  upper = rep(1, 3)
)

optimal_weights <- result$par / sum(result$par)
opt_return <- sum(optimal_weights * returns)
opt_risk   <- sqrt(t(optimal_weights) %*% covariance_matrix %*% optimal
    _weights)

cat("Optimal Weights (Max Sharpe):", round(optimal_weights, 4), "\n")

# ---- Plot Efficient Frontier + Max Sharpe Portfolio ----
plot(ef$Risk, ef$Return, type = "l", lwd = 2, col = "blue",
     xlab = "Portfolio Risk (Std Dev)",
     ylab = "Expected Return",
     main = "Efficient Frontier with Max Sharpe Portfolio")
points(ef$Risk, ef$Return, pch = 16, col = "blue")
points(opt_risk, opt_return, pch = 19, col = "red", cex = 1.5)
text(opt_risk, opt_return, labels = "Max Sharpe", pos = 4, col = "red")
```

```
Optimal Weights (Max Sharpe): 0.1675 0.3033 0.5291
```

As displayed in Figure 6.5-1, the blue curve represents the efficient frontier: the set of portfolios that provide the highest expected return for a given level of risk. Portfolios below this curve are inefficient because one could achieve higher returns at the same risk, or lower risk at the same return.

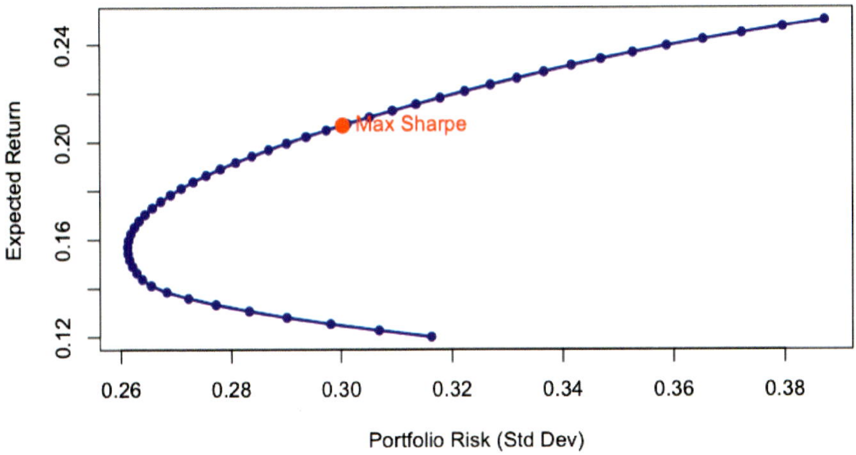

FIGURE 6.5-1: Efficient frontier of three assets, showing the trade-off between portfolio risk (standard deviation) and expected return. The red marker highlights the Max-Sharpe ratio portfolio, which delivers the highest risk-adjusted return among all feasible allocations.

The red point marks the Max-Sharpe portfolio. Positioned near the midsection of the frontier, it balances return and volatility to maximize risk-adjusted performance. Relative to other points on the frontier:

- Portfolios to the left (lower risk) sacrifice return more steeply.
- Portfolios to the right (higher return) demand disproportionate increases in risk.

Thus, the Max-Sharpe allocation is the most attractive for investors seeking optimal trade-offs between risk and return under these assumptions.

6.6 Case Study 5: Market-Making

Market-making is a cornerstone of financial markets, vital for sustaining liquidity and smooth trading. Traditionally, it was the domain of human traders who earned profits from the bid–ask spread by facilitating transactions. Their decisions drew on historical data, market indicators, and instinct, a method that worked reasonably well but often fell short in fast-moving and volatile markets.

The rise of AI has transformed market making, ushering in an era of greater efficiency and precision. Machine learning–driven algorithms can process vast datasets at speeds far beyond human capacity, enabling faster and more informed trading decisions. Through continuous learning and adaptation, these algorithms surpass traditional methods by uncovering subtle patterns and capitalizing on market inefficiencies.

The development, training, and real-time adaptation of AI models lie at the heart of modern market-making. Quantitative analysts and data scientists design algorithms to process live market data, training them on historical patterns so they can detect emerging trends and adjust rapidly to shifting conditions.

Consider a simplified example of an AI-driven market-making strategy implemented in R. For clarity, we focus on a basic algorithm built around a mean-reversion approach.

Listing 6.6.1: Mean-reverting market-making strategy with signals and plots.

```
library(dplyr)
library(zoo)
library(ggplot2)

set.seed(42)

# ---- Generate sample market data ----
timestamps <- seq.POSIXt(
  from = as.POSIXct("2023-01-01 00:00"),
  to   = as.POSIXct("2024-01-31 23:59"),
  by   = "min"
)

prices <- rnorm(length(timestamps), mean = 100, sd = 1)

market_data <- data.frame(
  Timestamp = timestamps,
  Price = prices
)

# ---- Define mean-reverting market-making strategy ----
mean_reverting_strategy <- function(data, window_size = 10, spread =
    0.1) {
  data <- data %>%
    mutate(
      RollingMean = rollmean(Price, k = window_size, fill = NA, align =
      "right"),
      UpperBound  = RollingMean + spread,
      LowerBound  = RollingMean - spread,
      Signal = case_when(
        Price < LowerBound ~ 1,    # Buy
        Price > UpperBound ~ -1,   # Sell
```

```
        TRUE ~ 0
      )
    )
  return(data)
}

# ---- Apply strategy ----
market_data <- mean_reverting_strategy(market_data)

# ---- Simulate trading positions ----
market_data <- market_data %>%
  mutate(Position = cumsum(Signal))

# ---- Plot strategy performance ----
ggplot(market_data, aes(x = Timestamp, y = Price)) +
  geom_line(aes(y = Price), color = "blue", alpha = 0.5) +
  geom_line(aes(y = RollingMean), color = "orange", linetype = "dashed"
    ) +
  geom_point(data = subset(market_data, Signal == 1),
             aes(x = Timestamp, y = Price),
             color = "green", shape = 24, size = 2, fill = "green") +
  geom_point(data = subset(market_data, Signal == -1),
             aes(x = Timestamp, y = Price),
             color = "red", shape = 25, size = 2, fill = "red") +
  labs(title = "Mean-Reverting Market-Making Strategy",
       x = "Timestamp", y = "Price") +
  theme_minimal()
```

The plot shown in Figure 6.6-1 illustrates how a mean-reverting market-making algorithm maintains a steady supply of bids (in green) and offers (in red) around a central mean price, which fluctuates roughly between 95 and 105. The profit mechanism is based on short-term price reversions rather than long-term forecasting. The significant clustering indicates frequent and continuous order posting throughout the year.

6.7 Case Study 6: Stock - Grouping

Hierarchical Cluster Analysis (HCA) groups stocks by similarity in a nested structure that uncovers relationships often missed by flat clustering methods. Unlike partitioning techniques such as *k-means*, which require the number of clusters to be set in advance, HCA builds a tree-like dendrogram that shows how stocks combine at varying levels of granularity. This makes HCA especially useful in finance, where sector overlaps, shifting market regimes, and latent factors create complex interdependencies.

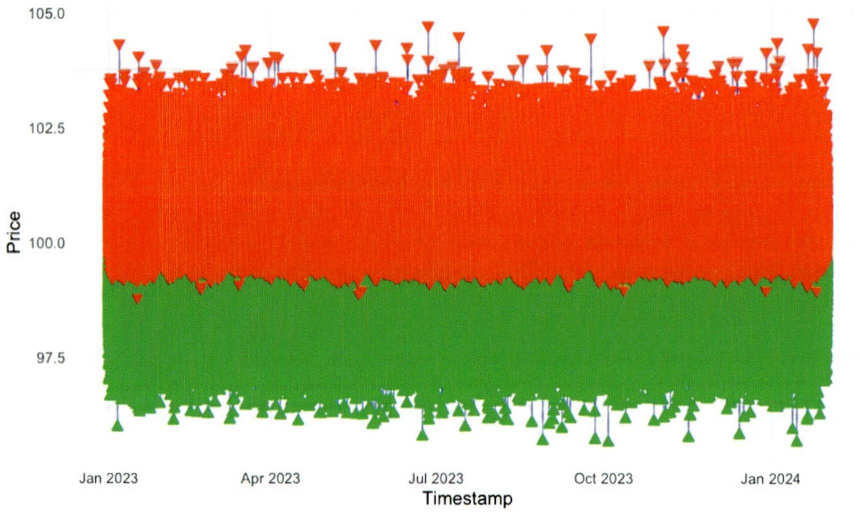

FIGURE 6.6-1: Mean-reverting market-making strategy with signals and plots.

Suppose you have daily return data for 200 stocks over several years. The goal is to identify natural groupings that reveal shared risk exposures or behavioral patterns. The process begins by defining a distance metric to measure similarity between stocks. While Euclidean distance on standardized returns is standard, correlation-based measures—such as one minus the correlation coefficient—are often preferred, as they better capture co-movement than raw price differences.

Listing 6.7.1 presents an example performing HCA on stock returns.

Listing 6.7.1: Hierarchical clustering dendrogram of stocks.

```
# ---- Packages ----
library(quantmod)
library(stats)
library(ggplot2)
library(reshape2)

# ---- Download data ----
tickers <- c("TSLA", "AAPL", "MSFT", "AMZN", "GOOG")
getSymbols(tickers, src = "yahoo",
           from = "2018-01-01", to = "2023-12-31",
           auto.assign = TRUE)

prices <- do.call(merge, lapply(tickers, function(t) Ad(get(t))))
colnames(prices) <- tickers
```

```r
returns <- na.omit(ROC(prices, type = "discrete"))
returns_mat <- as.matrix(returns)

# ---- Calculate the correlation matrix between stocks ----
corr_matrix <- cor(returns, use = "pairwise.complete.obs")

# ---- Convert correlation to distance: distance = 1 - correlation ----
distance_matrix <- as.dist(1 - corr_matrix)

# ---- Perform hierarchical clustering using Ward's method ----
hc <- hclust(distance_matrix, method = "ward.D2")

# ---- Plot dendrogram to visualize cluster hierarchy ----
plot(hc,
     main = "Hierarchical Clustering Dendrogram of Stocks",
     xlab = "Stock Ticker",
     ylab = "Distance",
     sub = "",
     hang = -1,
     cex = 0.6)

# Add rectangles around clusters (optional)
rect.hclust(hc, h = 0.7 * max(hc$height), border = "red")
```

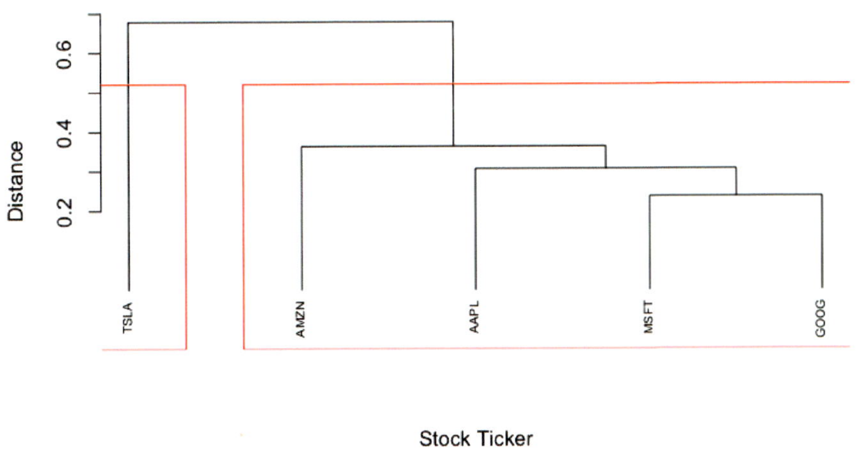

FIGURE 6.7-1: Hierarchical clustering dendrogram of selected technology stocks (TSLA, AMZN, AAPL, MSFT, GOOG) based on daily return correlations. The red horizontal line indicates the distance threshold used to define clusters.

The dendrogram displayed in Figure 6.7-1 reveals how stocks group together according to the similarity of their return patterns. At lower distances,

MSFT and GOOG cluster first, reflecting their highly correlated price movements as large-cap tech peers. AAPL joins next, forming a broader technology cluster. AMZN links to this group at a slightly higher distance, suggesting weaker—but still significant—co-movement with the others. TSLA remains separate until the highest distance threshold, indicating that its return dynamics are less aligned with the other tech firms, perhaps due to its unique business model and higher idiosyncratic volatility. Overall, the dendrogram highlights natural sub-groups within the tech sector, with MSFT–GOOG–AAPL forming the core cluster and TSLA showing distinct behavior.

In R, `cutree()` slices the dendrogram at a specified height. The code below applies a threshold equal to 70% of the maximum merge height, assigns each stock to a cluster, and builds a tidy DataFrame mapping tickers to their cluster IDs. Finally, it sorts and displays the first ten entries for inspection.

Listing 6.7.2: `cutree()` to slice the dendrogram at a specified height.

```
# ---- Packages ----
library(stats)
library(dplyr)

# ---- Assume 'linkage_matrix' is created using hclust ----
# For example:
dist_matrix <- dist(t(returns))          # distance between stock
    returns
linkage_matrix <- hclust(dist_matrix, method = "ward.D2")

# ---- Extract clusters with max distance threshold ----
max_distance <- 0.7 * max(linkage_matrix$height)

clusters <- cutree(linkage_matrix, h = max_distance)

# ---- Create DataFrame mapping tickers to cluster IDs ----
cluster_df <- data.frame(
  Ticker = colnames(returns),
  Cluster = clusters
)

# ---- Display first 10 tickers sorted by cluster ----
cluster_df %>%
  arrange(Cluster) %>%
  head(10)
```

```
      Ticker Cluster
TSLA    TSLA       1
AAPL    AAPL       2
MSFT    MSFT       2
```

```
 AMZN   AMZN       2
 GOOG   GOOG       2
```

This output enables targeted analysis within clusters—for example, calculating average sector weights or examining intra-cluster correlations. Some clusters align neatly with familiar sectors such as financials or consumer staples, while others uncover hidden structures shaped by volatility regimes, geographic exposure, or other latent factors.

Hierarchical clustering enhances portfolio construction and risk management by identifying groups of assets that may experience joint drawdowns due to shared influences, like economic cycles. This insight aids in developing more effective hedging strategies. Ward's linkage method minimizes within-cluster variance to produce clear and interpretable groups. Alternatively, methods like single or complete linkage can reveal different market patterns. However, hierarchical clustering assumes stable asset relationships throughout the analysis period, which is often not the case in financial markets.

Gaussian Mixture Models

Gaussian Mixture Models (GMMs) offer a probabilistic approach to clustering that overcomes some drawbacks of *k-means* and *hierarchical* methods. Instead of assigning each observation to a single cluster, GMM models the data as arising from a mixture of Gaussian distributions, enabling "soft" clustering where points can belong to multiple clusters with varying probabilities. This makes GMM particularly effective in financial contexts, where cluster boundaries are rarely clear-cut.

Applied to stock returns, GMM can uncover nuanced structures such as assets that straddle sectors or shift with changing regimes. By incorporating covariance, GMM produces elliptical clusters that more closely match real-world return distributions, helping reveal asymmetric risk profiles and correlated behaviors. For illustration, suppose we have a DataFrame returns containing daily returns for 100 stocks over several years. Using **mclust**, which implements Gaussian finite mixture modeling using the EM algorithm, we can fit a GMM to detect latent factors driving market dynamics.

Listing 6.7.3: Gaussian Mixture Model.

```
# ---- Packages ----
library(mclust)

# ---- Example: Simulated financial returns ----
set.seed(123)
returns <- data.frame(
  stock1 = rnorm(200, mean = 0.001, sd = 0.02),
  stock2 = rnorm(200, mean = -0.0005, sd = 0.015)
)
```

```r
# ---- Fit Gaussian Mixture Model ----
gmm_fit <- Mclust(returns, G = 1:5)  # test 1-5 mixture components

# ---- Summary ----
summary(gmm_fit)

# ---- Cluster assignments ----
head(gmm_fit$classification)

# ---- Plot (optional) ----
plot(gmm_fit, what = "classification")
```

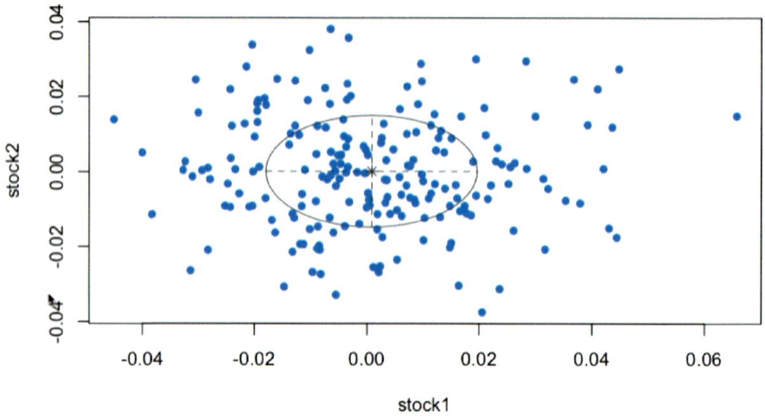

FIGURE 6.7-2: Scatterplot of daily returns for two stocks with Gaussian mixture ellipse capturing the central distribution.

The scatterplot displayed in Figure 6.7-2 shows the joint distribution of daily returns for stock1 and stock2. Each point represents a trading day. The fitted ellipse indicates the estimated Gaussian component from the mixture model, capturing the region where most return observations cluster. Both stocks exhibit returns concentrated around zero, with some dispersion reflecting market volatility. The elliptical shape suggests mild positive correlation, meaning that when one stock experiences positive returns, the other tends to move in the same direction. Outliers outside the ellipse highlight days with atypical co-movement, potentially driven by market shocks or stock-specific events.

6.8 Case Study 7: Predicting Stock Trends

Predicting stock trends remains one of the core challenges in financial machine learning, demanding both market intuition and data-driven rigor. While no model can guarantee perfect foresight in the face of noisy and multifaceted markets, structured approaches that emphasize careful feature engineering, thoughtful model selection, and systematic tuning can meaningfully improve predictive accuracy.

For illustration, consider a mid-cap technology stock with moderately volatile daily price movements, often shaped by earnings announcements and macroeconomic events. We will develop a pipeline that leverages historical price data, engineered features, and supervised learning methods to forecast whether the stock's closing price will rise or fall the following day.

The dataset includes daily OHLCV (Open, High, Low, Close, Volume) data spanning two years. The first step is to create meaningful input variables beyond raw prices that capture relevant market signals. Common technical indicators often provide additional insight.

- **Moving Averages (MA):** Short-term fluctuations. For example, a 10-day simple moving average (SMA) helps reveal trend direction.
- **Relative Strength Index (RSI):** Measures momentum to identify overbought or oversold conditions.
- **Moving Average Convergence Divergence:** Tracks momentum shifts by comparing short-term and long-term exponential moving averages (EMAs).
- **Volume Weighted Average Price (VWAP):** Represents the average trading price weighted by volume, often signaling institutional activity.

Listing 6.8.1: Stock trends using Technical Trading Rules.

```
# ---- Packages ----
library(quantmod)
library(TTR)
library(dplyr)

# ---- Download stock data (e.g., Tesla TSLA) ----
ticker <- "TSLA"
getSymbols(ticker, src = "yahoo", from = "2020-01-01", to = "2025-12-31
    ", auto.assign = TRUE)

# Extract OHLCV data
data_xts <- get(ticker)
```

```
# Convert to data frame
data <- data.frame(
  Date   = index(data_xts),
  Open   = as.numeric(Op(data_xts)),
  High   = as.numeric(Hi(data_xts)),
  Low    = as.numeric(Lo(data_xts)),
  Close  = as.numeric(Cl(data_xts)),
  Volume = as.numeric(Vo(data_xts))
)

# Preview
head(data)

# Convert Date column to Date type
data$Date <- as.Date(data$Date)

# ---- Calculate indicators ----
# 10-day Simple Moving Average
data$SMA_10 <- SMA(data$Close, n = 10)

# RSI over 14 days
data$RSI_14 <- RSI(data$Close, n = 14)

# MACD components
macd_vals <- MACD(data$Close, nFast = 12, nSlow = 26, nSig = 9)
data$MACD <- macd_vals[, 1]          # MACD line
data$MACD_signal <- macd_vals[, 2]   # Signal line

# VWAP approximation using Close * Volume / cumVolume
data$VWAP <- cumsum(data$Close * data$Volume) / cumsum(data$Volume)

# ---- Drop rows with NA values (from indicators) ----
data <- na.omit(data)

# ---- Create binary target ----
# 1 if next day's Close > today's Close, else 0
data$Target <- ifelse(dplyr::lead(data$Close) > data$Close, 1, 0)

# ---- Drop last row (no future close available) ----
data <- data[1:(nrow(data)-1), ]

# ---- Preview final dataset ----
head(data)
```

```
         Date     Open     High      Low    Close    Volume   SMA_10
34 2020-02-20 60.79667 60.80000 57.32933 59.96067 264523500 53.92827
```

```
35 2020-02-21 60.46533 60.87067 58.69667 60.06667 214722000 54.94187
36 2020-02-24 55.93333 57.56667 54.81333 55.58600 227883000 55.51333
      RSI_14   MACD  MACD_signal VWAP     Target
34    74.64493 15.48510 16.58099  44.16879 1
35    74.74181 15.13806 16.29241  44.45879 0
36    63.66776 14.01859  15.83764 44.67012 0
```

By overlaying the binary target directly onto the price chart (Figure 6.8-1), we can visualize how predictive signals align with actual market behavior. Green dots mark periods where the model anticipated the next day's close to be higher than today's (Target = 1), while red dots highlight days where the next session was expected to close lower (Target = 0). This color-coded overlay transforms the chart into a more intuitive decision aid: instead of viewing price, SMA, and VWAP in isolation, traders can immediately see the directional bias implied by the target variable. Such a visualization bridges raw technical indicators with predictive modeling outcomes, providing a clearer link between signal generation and actionable insights.

Moreover, Tesla's price dynamics, when analyzed with technical overlays (Figure 6.8-1), reveal distinct cycles of momentum, exhaustion, and recovery. The stock's rapid appreciation into late 2021 gave way to a steep decline through 2022 before settling into a stabilizing range in 2023. The 10-day SMA tracked short-term swings closely, while the VWAP, anchored to the longer-term average purchase price, consistently lagged nearer to the underlying trend baseline. Crossings of price below the VWAP underscored sustained periods of weakness during the downturn. Momentum indicators echoed these phases: the RSI(14) oscillated within the 30–70 band, with repeated breaches above 70 in 2020–2021 marking overbought conditions amid Tesla's rally, and dips below 30 in mid-2022 signaling oversold episodes as selling pressure intensified. Meanwhile, the MACD fluctuated around the zero line, with bullish crossovers in late 2020 and early 2023 highlighting momentum recoveries, while the frequent bearish crossovers of 2022 reinforced the persistence of downward pressure. Together, these measures provide a layered view of Tesla's shifting market regime.

6.9 Case Study 8: PCA Application

Suppose we analyze daily returns for 500 stocks over several years. Because of shared sector influences, macroeconomic forces, and broad market movements, these returns tend to be correlated. Feeding all such raw features directly into a predictive model risks overfitting and excessive computational demands. Principal Component Analysis (PCA) mitigates these issues by extracting

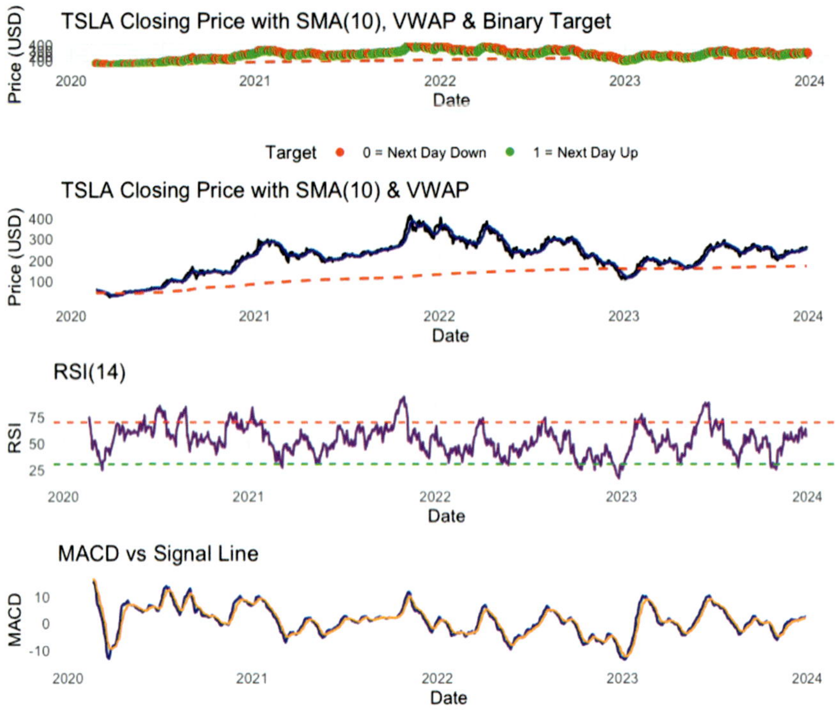

FIGURE 6.8-1: Tesla (TSLA) technical indicators with predictive target overlay. The top panel shows closing price with SMA(10) and VWAP, where binary targets are marked as green dots (Target = 1, next day's close > today's close) and red dots (Target = 0, next day's close ≤ today's close). The middle panel displays RSI(14) with standard overbought (70) and oversold (30) thresholds, while the bottom panel presents MACD and, its signal line.

principal components—linear combinations of the original returns—that capture the dominant market dynamics. In practice, only a handful of components often explain most of the variance, reducing complexity and making the relationships more tractable.

Listing 6.9.1: Principal Component Analysis (PCA) application.

```
# ---- Packages ----
library(readr)
library(ggplot2)

# ---- Packages ----
library(quantmod)

# ---- Example: Download multiple stocks ----
```

```r
tickers <- c("AAPL", "MSFT", "GOOG", "AMZN", "TSLA")
getSymbols(tickers, src = "yahoo",
           from = "2020-01-01", to = "2023-12-31",
           auto.assign = TRUE)

# ---- Extract adjusted prices ----
prices <- do.call(merge, lapply(tickers, function(t) Ad(get(t))))
colnames(prices) <- tickers

# ---- Compute daily returns ----
returns <- na.omit(ROC(prices, type = "discrete"))

# ---- Standardize returns (mean=0, variance=1) ----
returns_std <- scale(returns)

# Quick check
head(returns_std)

# ---- PCA: extract first 5 components ----
pca <- prcomp(returns_std, center = TRUE, scale. = TRUE)

# Explained variance ratio
explained_var <- (pca$sdev^2) / sum(pca$sdev^2)
cat("Explained variance ratio:\n")
print(explained_var[1:5])

# ---- Cumulative explained variance ----
cum_explained_var <- cumsum(explained_var)

# ---- Plot cumulative explained variance ----
df <- data.frame(
  PC = 1:length(cum_explained_var),
  CumulativeVariance = cum_explained_var
)

ggplot(df[1:5, ], aes(x = PC, y = CumulativeVariance)) +
  geom_line(color = "blue") +
  geom_point(shape = 21, size = 3, fill = "red") +
  labs(
    title = "PCA Explained Variance",
    x = "Number of Principal Components",
    y = "Cumulative Explained Variance"
  ) +
  theme_minimal()
```

```
Explained variance ratio:
[1] 0.69212551 0.13134614 0.07885730 0.05822720 0.03944385
```

The explained variance ratio quantifies how much of the total variability in the dataset each principal component captures. Here, the first component alone accounts for about 69.2 percent of the variance, indicating that a single dominant factor drives most of the co-movement across the 500 stocks—likely reflecting broad market or macroeconomic influences. The second component contributes another 13.1 percent, and the third adds 7.9 percent, together explaining over 90 percent of the variance. By the time we include the fourth (5.8 percent) and fifth (3.9 percent) components, the cumulative variance explained rises to nearly 99 percent.

This distribution suggests that the high-dimensional return space can be effectively reduced to a handful of components without losing much information. The first principal component captures the "market mode," while subsequent components likely represent sectoral or style-based effects (e.g., technology vs. utilities, growth vs. value). In practical terms, using the top three to five components provides a compact yet powerful representation of the underlying market structure, mitigating noise and overfitting risks.

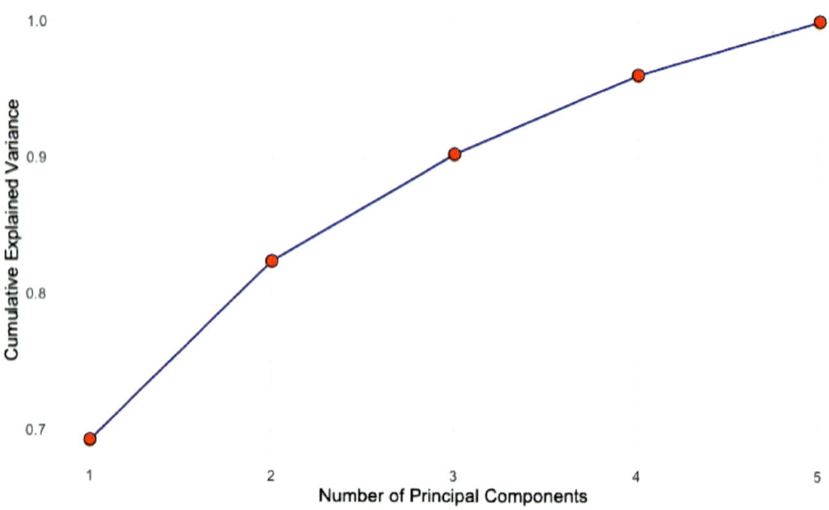

FIGURE 6.9-1: Cumulative explained variance of the first five principal components. The curve shows how successive components account for an increasing share of the total variance in stock returns.

The plot in Figure 6.9-1 illustrates that the first principal component alone explains nearly 70 percent of the total variance, capturing the dominant market-wide factor. Adding the second component raises the cumulative explained variance above 80 percent, while the third brings it to around 90 percent. By the fourth and fifth components, nearly all the variance is accounted for. This pattern confirms that the high-dimensional return data can be effectively summarized with just a few principal components, highlighting the strong common structure in equity returns and the potential for dimensionality reduction in predictive modeling.

6.10 Case Study 9: Unsupervised Portfolio Analysis

Exploring unsupervised learning in portfolio analysis reveals how subtle patterns emerge without explicit labels. Consider a mid-sized investment firm aiming to enhance diversification by uncovering hidden groupings among 150 stocks across multiple sectors. Traditional classifications often overlook evolving correlations and latent dependencies, making unsupervised methods especially valuable. Using three years of daily returns, the firm begins by standardizing the data to remove scale effects, a standard step in time-series preprocessing. PCA then reduces the dimensionality from 150 stocks to five principal components, capturing about 80 percent of the variance. This step filters out noise while retaining the dominant market drivers. Building on this compressed representation, GMMs identify overlapping clusters that reflect latent market regimes or asset classes beyond conventional GICS sectors.

In R, the Gaussian Mixture with BIC-based model selection are provided by **mclust** and **Rmixmod** packages. The most common is **mclust**.

Listing 6.10.1: **mclust** Gaussian Mixture Modeling.

```r
# ---- Packages ----
library(mclust)

# Example: returns matrix of stock data (rows = obs, cols = assets)
set.seed(123)
data <- matrix(rnorm(300), ncol = 3)

# ---- Fit Gaussian Mixture Models with BIC ----
gmm_fit <- Mclust(data)

# ---- Optimal number of clusters ----
summary(gmm_fit)

# BIC values for each model
plot(gmm_fit, what = "BIC")

# Cluster assignments
clusters <- gmm_fit$classification
head(clusters)
```

TABLE 6.5: Model fit summary with log–likelihood, sample size (n), degrees of freedom (df), Bayesian Information Criterion (BIC), and Integrated Completed Likelihood (ICL).

log–likelihood	n	df	BIC	ICL
-406.7257	100	4	-831.8721	-831.8721

Rmixmod

Rmixmod offers an R interface to the MIXMOD software library, allowing users to conduct model-based mixture modeling for unsupervised clustering, as well as supervised and semi-supervised classification. It enables the analysis of multivariate Gaussian and multinomial mixture models to estimate densities, classify data, and identify clusters within datasets.

Listing 6.10.2: **Rmixmod** Gaussian Mixture Modeling.

```
library(Rmixmod)

# Fit models for 1 to 6 clusters
res <- mixmodCluster(data_df, nbCluster = 1:6)

# Print summary
res

# Extract the best model directly
best_model <- res@bestResult

# Inspect details
best_model@nbCluster        # optimal number of clusters
best_model@criterionValue   # BIC value
best_model@partition        # cluster assignments
```

```
*** INPUT:
****************************************
* nbCluster =  1 2 3 4 5 6
* criterion =  BIC
****************************************
*** MIXMOD Models:
* list =  Gaussian_pk_Lk_C
* This list includes only models with free proportions.
****************************************
* data (limited to a 10x10 matrix) =
        V1      V2       V3
  [1,] -0.5605 -0.7104  2.199
```

```
  [2,] -0.2302 0.2569   1.312
  [3,] 1.559   -0.2467  -0.2651
  [4,] 0.07051 -0.3475  0.5432
  [5,] 0.1293  -0.9516  -0.4143
  [6,] 1.715   -0.04503 -0.4762
  [7,] 0.4609  -0.7849  -0.7886
  [8,] -1.265  -1.668   -0.5946
  [9,] -0.6869 -0.3802  1.651
 [10,] -0.4457 0.919    -0.05403
 * ... ...
 ****************************************
 *** MIXMOD Strategy:
 * algorithm              =  EM
 * number of tries        =  1
 * number of iterations =  200
 * epsilon                =  0.001
 *** Initialization strategy:
 * algorithm              =  smallEM
 * number of tries        =  10
 * number of iterations =  5
 * epsilon                =  0.001
 * seed                   =  NULL
 ****************************************

 ****************************************
 *** BEST MODEL OUTPUT:
 *** According to the BIC criterion
 ****************************************
 * nbCluster   =  1
 * model name  =  Gaussian_pk_Lk_C
 * criterion   =  BIC(852.5641)
 * likelihood  =  -405.5588
 ****************************************
 *** Cluster 1
 * proportion =  1.0000
 * means      =  0.0904 -0.1075 0.1205
 * variances  = |     0.8249    -0.0433    -0.1109 |
                |    -0.0433     0.9257     0.0278 |
                |    -0.1109     0.0278     0.8932 |
 ****************************************
```

The variance–covariance matrix suggests the three variables are moderately correlated (non-zero off-diagonal entries), and each has variance around 0.9.

Listing 6.10.3: BIC vs. number of clusters plot for **Rmixmod**.

```
library(Rmixmod)
library(ggplot2)

# ---- Fit models for 1 to 6 clusters ----
res <- mixmodCluster(data_df, nbCluster = 1:6)

# ---- Extract BIC values into a data.frame ----
bic_df <- data.frame(
  k   = sapply(res@results, function(r) r@nbCluster),
  BIC = sapply(res@results, function(r) r@criterionValue)
)

# ---- ggplot2 visualization ----
ggplot(bic_df, aes(x = k, y = BIC)) +
  geom_line(color = "steelblue", linewidth = 1) +
  geom_point(color = "darkred", size = 3) +
  theme_minimal(base_size = 14) +
  labs(title = "BIC vs Number of Clusters",
       x = "Number of Clusters (k)",
       y = "BIC") +
  theme(plot.title = element_text(hjust = 0.5, face = "bold"))
```

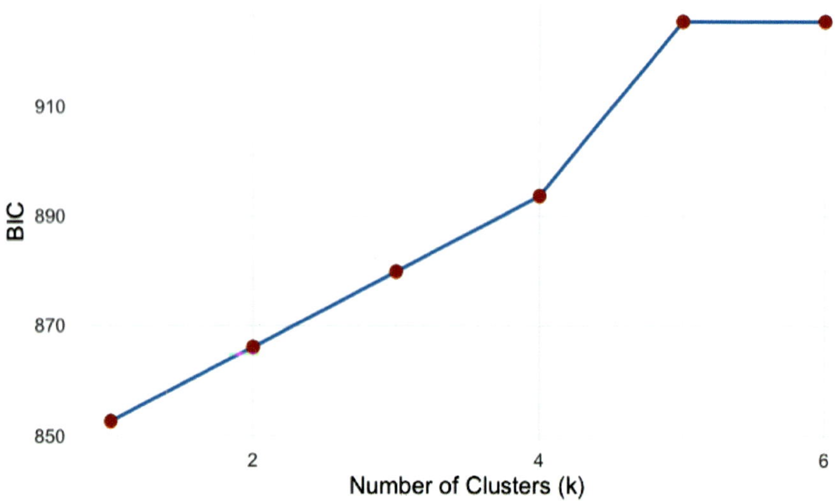

FIGURE 6.10-1: The plot shows the Bayesian Information Criterion (BIC) values across candidate cluster counts $k = 1$ through 6.

The BIC analysis displayed in Figure 6.10-1 indicates that the best-fitting model occurs at $k = 1$, where the criterion reaches its lowest value. Because a lower BIC reflects a superior balance between goodness of fit and model complexity, this result suggests that the data are best described as a single cluster rather than multiple distinct groups. As the number of clusters increases, the BIC rises consistently, showing that the added complexity of additional clusters does not yield sufficient improvement in fit to justify their inclusion. In practical terms, this outcome implies that the dataset is relatively homogeneous, and the Gaussian mixture model does not uncover any meaningful substructure—making the introduction of multiple clusters unnecessary.

6.11 Case Study 10: Deep Learning Models

Reinforcement learning (RL) differs from traditional supervised and unsupervised methods by learning through direct interaction with an environment. This makes it well suited to trading, where decisions occur sequentially and must adapt to shifting market conditions. Unlike static prediction models, an RL agent updates its behavior continually, guided by rewards or penalties from its trading actions, with the goal of maximizing cumulative returns. This dynamic process mirrors the uncertainty and evolving character of financial markets.

To illustrate, consider a simplified RL agent operating on daily stock price data. Each day, it decides whether to buy, hold, or sell shares based on inputs such as current price and technical indicators. Profitable trades yield positive rewards proportional to gains, while losses generate negative feedback. Over time, algorithms like Q-learning or policy gradients refine the agent's strategy by detecting recurring patterns that lead to better outcomes.

Building an RL trading system requires defining several core components:

- **State Representation:** Capturing relevant market information efficiently—such as recent price history, volume, volatility measures, or derived features like moving averages—to inform decisions.
- **Action Space:** Specifying allowable actions (e.g., buy, sell, hold) along with any constraints such as position limits or transaction costs.
- **Reward Function:** Converting trading results into numerical feedback signals—commonly daily profit loss adjusted for risk exposure—that guide learning.
- **Policy Model:** The algorithmic framework that directs action selection; modern systems often use neural networks to approximate this policy.

Listing 6.11.1: Reinforcement learning I.

```
library(reticulate)

# Make sure Python deps are installed (only once if needed)
# py_install(c("stable-baselines3", "gym", "yfinance", "numpy", "pandas
    "), pip = TRUE)

py_run_string("
import gym
import numpy as np
import pandas as pd
import yfinance as yf
from stable_baselines3 import PPO
from stable_baselines3.common.vec_env import DummyVecEnv

# ---- Custom environment ----
class StockTradingEnv(gym.Env):
    def __init__(self, df):
        super(StockTradingEnv, self).__init__()
        self.df = df.reset_index(drop=True)
        self.current_step = 0
        self.initial_balance = 10000
        self.balance = self.initial_balance
        self.shares_held = 0
        self.net_worth = self.initial_balance

        # Actions: 0=hold, 1=buy, 2=sell
        self.action_space = gym.spaces.Discrete(3)

        # Observations: OHLC + shares held
        self.observation_space = gym.spaces.Box(
            low=0, high=np.inf, shape=(5,), dtype=np.float32
        )

    def _next_observation(self):
        frame = self.df.iloc[self.current_step]
        return np.array([
            float(frame['Open']),
            float(frame['High']),
            float(frame['Low']),
            float(frame['Close']),
            float(self.shares_held)
        ], dtype=np.float32)

    def _take_action(self, action):
        current_price = float(self.df.iloc[self.current_step]['Close'])
        if action == 1 and self.balance >= current_price:
            self.shares_held += 1
            self.balance -= current_price
```

```
        elif action == 2 and self.shares_held > 0:
            self.shares_held -= 1
            self.balance += current_price

    def step(self, action):
        self._take_action(action)
        self.current_step += 1
        done = self.current_step >= (len(self.df) - 1)
        current_price = float(self.df.iloc[self.current_step]['Close'])
        self.net_worth = self.balance + (self.shares_held * current_
    price)
        reward = self.net_worth - self.initial_balance
        return self._next_observation(), reward, done, {}

    def reset(self):
        self.balance = self.initial_balance
        self.shares_held = 0
        self.net_worth = self.initial_balance
        self.current_step = 0
        return self._next_observation()

# ---- Download data ----
df = yf.download('AAPL', start='2018-01-01', end='2025-01-01')[['Open
    ','High','Low','Close']].reset_index()

# ---- Wrap environment ----
env = DummyVecEnv([lambda: StockTradingEnv(df)])

# ---- Train PPO agent ----
model = PPO('MlpPolicy', env, verbose=1)
model.learn(total_timesteps=50000)

# ---- Evaluate ----
obs = env.reset()
for _ in range(len(df)-1):
    action, _states = model.predict(obs)
    obs, reward, done, info = env.step(action)
    if done:
        break

final_value = env.envs[0].net_worth
")

# ---- Access in R ----
cat('Final portfolio value:', py$final_value, '\n')
```

```
-----------------------------
| time/              |      |
|    fps             | 1664 |
|    iterations      | 1    |
|    time_elapsed    | 1    |
|    total_timesteps | 2048 |
-----------------------------
-------------------------------------------
| time/                   |               |
|    fps                  | 1369          |
|    iterations           | 2             |
|    time_elapsed         | 2             |
|    total_timesteps      | 4096          |
| train/                  |               |
|    approx_kl            | 0.00023649336 |
|    clip_fraction        | 0             |
|    clip_range           | 0.2           |
|    entropy_loss         | -1.1          |
|    explained_variance   | 2.21e-06      |
|    learning_rate        | 0.0003        |
|    loss                 | 1.67e+08      |
|    n_updates            | 10            |
|    policy_gradient_loss | -0.000534     |
|    value_loss           | 3.02e+08      |
-------------------------------------------
...
-------------------------------------------
| time/                   |               |
|    fps                  | 1297          |
|    iterations           | 24            |
|    time_elapsed         | 37            |
|    total_timesteps      | 49152         |
| train/                  |               |
|    approx_kl            | 4.4275657e-07 |
|    clip_fraction        | 0             |
|    clip_range           | 0.2           |
|    entropy_loss         | -1.09         |
|    explained_variance   | 3.66e-05      |
|    learning_rate        | 0.0003        |
|    loss                 | 5.26e+09      |
|    n_updates            | 230           |
|    policy_gradient_loss | -7.75e-06     |
|    value_loss           | 1.41e+10      |
-------------------------------------------
------------------------------------------
| time/                   |              |
|    fps                  | 1297         |
```

```
|    iterations           | 25           |
|    time_elapsed         | 39           |
|    total_timesteps      | 51200        |
| train/                  |              |
|    approx_kl            | 3.202731e-06 |
|    clip_fraction        | 0            |
|    clip_range           | 0.2          |
|    entropy_loss         | -1.09        |
|    explained_variance   | 5.96e-08     |
|    learning_rate        | 0.0003       |
|    loss                 | 6.37e+09     |
|    n_updates            | 240          |
|    policy_gradient_loss | -5e-05       |
|    value_loss           | 1.35e+10     |
------------------------------------------
Final portfolio value: 10000
```

Interpretation–Key Sections in the Log

1. **time-block**

   - `fps`: how many environment steps per second are processed: (1300–1600 means your env is running efficiently).
   - `iterations`: PPO update cycles (each corresponds to collecting n_steps timesteps, here 2048 per rollout).
   - `total_timesteps`: total env steps experienced by the agent.
   - `time_elapsed`: wall-clock seconds since training started. → These tell you training speed and how far along the rollout is.

2. **train-block**

   - `approx_kl`: the KL-divergence between old and new policy. A small number (1e-4 to 1e-3) is healthy. If it grows large, the policy update is diverging.
   - `clip_fraction`: fraction of actions where PPO clipping activated. At 0, the policy updates are small (not bumping into the clipping threshold).
   - `clip_range`: the configured PPO clipping range (default 0.2).
   - `entropy_loss`: negative entropy of the policy. Around -1.1 is constant here, meaning the action distribution has stable randomness (policy not collapsing).
   - `explained_variance`: measures how well the value function predicts returns. Near 0 means the critic isn't learning useful signal yet.

- `learning_rate`: optimizer step size (default 0.0003).
- `loss`: total loss (policy + value). Large values (1e8–1e10) reflect scale, not necessarily a problem.
- `policy_gradient_loss`: contribution of the policy update. Negative and very small here, consistent with minor updates.
- `value_loss`: critic's regression error. Huge and noisy (1e8–1e10), suggesting the value function is struggling with scale.

3. **Interpretation**

   - `Training is progressing`: PPO is rolling through iterations, timesteps increasing in chunks of 2048 as expected.
   - `Speed is good`: 1300 FPS shows the environment is lightweight.
   - `Policy updates are tiny`: clip_fraction = 0 and very small policy_gradient_loss → the agent is barely adjusting its behavior.
   - `Critic (value function) unstable`: explained_variance ≈ 0, value_loss very large → the model isn't learning to predict returns properly.
   - `Entropy steady`: the agent is maintaining some exploration.

4. **Significance**
   Right now, the PPO agent is not really learning anything useful in your environment:

   - The critic can't explain returns (`explained_variance ≈ 0`).
   - Value loss is exploding in scale.
   - Policy updates are minuscule.

   This often happens if:

   - Rewards are scaled very large (here reward = net worth - 10000, which grows fast).
   - Observations are not normalized.
   - The environment's step/reward structure doesn't give dense feedback.

5. **Next steps**:

   - Normalize rewards (e.g., reward = % change in net worth, not absolute dollar change).
   - Scale observations (divide OHLC by a baseline, like starting price).
   - Use `SB3's VecNormalize` wrapper to handle normalization automatically.
   - Train longer (50 k timesteps is very small for PPO).

Rewrite the custom `StockTradingEnv` (Listing 6.11.1) so that rewards are based on percentage returns (per-step change in net worth), instead of raw absolute net worth. This stabilizes learning for PPO. Here's the full R script with embedded Python:

Listing 6.11.2: Re-enforcement learning II.

```
library(reticulate)

# Ensure Python deps installed (only once if needed)
# py_install(c("stable-baselines3", "gym", "yfinance", "numpy", "pandas
    "), pip = TRUE)

py_run_string("
import gym
import numpy as np
import pandas as pd
import yfinance as yf
from stable_baselines3 import PPO
from stable_baselines3.common.vec_env import DummyVecEnv, VecNormalize

# ---- Custom environment ----
class StockTradingEnv(gym.Env):
    def __init__(self, df):
        super(StockTradingEnv, self).__init__()
        self.df = df.reset_index(drop=True)
        self.current_step = 0
        self.initial_balance = 10000
        self.balance = self.initial_balance
        self.shares_held = 0
        self.net_worth = self.initial_balance

        # Actions: 0=hold, 1=buy, 2=sell
        self.action_space = gym.spaces.Discrete(3)

        # Observations: OHLC + shares held
        self.observation_space = gym.spaces.Box(
            low=0, high=np.inf, shape=(5,), dtype=np.float32
        )

    def _next_observation(self):
        frame = self.df.iloc[self.current_step]
        return np.array([
            float(frame['Open']),
            float(frame['High']),
            float(frame['Low']),
            float(frame['Close']),
            float(self.shares_held)
        ], dtype=np.float32)
```

```
    def _take_action(self, action):
        current_price = float(self.df.iloc[self.current_step]['Close'])
        if action == 1 and self.balance >= current_price:
            self.shares_held += 1
            self.balance -= current_price
        elif action == 2 and self.shares_held > 0:
            self.shares_held -= 1
            self.balance += current_price

    def step(self, action):
        prev_net_worth = self.net_worth

        self._take_action(action)
        self.current_step += 1
        done = self.current_step >= (len(self.df) - 1)

        current_price = float(self.df.iloc[self.current_step]['Close'])
        self.net_worth = self.balance + (self.shares_held * current_
    price)

        # Reward = percentage change in net worth
        reward = (self.net_worth - prev_net_worth) / prev_net_worth

        return self._next_observation(), reward, done, {}

    def reset(self):
        self.balance = self.initial_balance
        self.shares_held = 0
        self.net_worth = self.initial_balance
        self.current_step = 0
        return self._next_observation()

# ---- Download data ----
df = yf.download('AAPL', start='2018-01-01', end='2025-01-01')[['Open
    ','High','Low','Close']].reset_index()

# ---- Wrap environment with normalization ----
raw_env = DummyVecEnv([lambda: StockTradingEnv(df)])
env = VecNormalize(raw_env, norm_obs=True, norm_reward=True, clip_obs
    =10.0)

# ---- Train PPO agent ----
model = PPO('MlpPolicy', env, verbose=1)
model.learn(total_timesteps=50000)

# ---- Evaluate ----
obs = env.reset()
for _ in range(len(df)-1):
    action, _states = model.predict(obs)
```

```
    obs, reward, done, info = env.step(action)
    if done:
        break

final_value = env.venv.envs[0].net_worth
")

# ---- Access in R ----
cat('Final portfolio value:', py$final_value, '\n')
```

```
----------------------------
| time/              |      |
|    fps             | 1517 |
|    iterations      | 1    |
|    time_elapsed    | 1    |
|    total_timesteps | 2048 |
----------------------------
-----------------------------------------
| time/                   |             |
|    fps                  | 1267        |
|    iterations           | 2           |
|    time_elapsed         | 3           |
|    total_timesteps      | 4096        |
| train/                  |             |
|    approx_kl            | 0.00936123  |
|    clip_fraction        | 0.0455      |
|    clip_range           | 0.2         |
|    entropy_loss         | -1.09       |
|    explained_variance   | 0.0299      |
|    learning_rate        | 0.0003      |
|    loss                 | 0.297       |
|    n_updates            | 10          |
|    policy_gradient_loss | -0.00331    |
|    value_loss           | 0.602       |
-----------------------------------------
...
-----------------------------------------
| time/                   |             |
|    fps                  | 1229        |
|    iterations           | 25          |
|    time_elapsed         | 41          |
|    total_timesteps      | 51200       |
| train/                  |             |
|    approx_kl            | 0.011607178 |
|    clip_fraction        | 0.0655      |
|    clip_range           | 0.2         |
|    entropy_loss         | -0.96       |
```

```
|    explained_variance    | 0.801      |
|    learning_rate         | 0.0003     |
|    loss                  | 0.0339     |
|    n_updates             | 240        |
|    policy_gradient_loss  | -0.00334   |
|    value_loss            | 0.0919     |
-----------------------------------------
```

This new log shows a big improvement compared to the earlier runs with absolute net worth rewards.

1. **Explained Variance**

   - `Before (absolute rewards)`: ≈0.0 throughout → critic (value function) wasn't learning at all.
   - `Now (percentage rewards)`: starts ≈0.03 and climbs steadily → 0.80 by 50 k timesteps.

     ⋆ *This means the critic is now able to explain ≈80 percent of return variance → strong signal of learning.*

2. **Value Loss**

   - **Before**: gigantic values (1e8 to 1e10) → critic exploding due to unscaled rewards.
   - **Now**: small and stable (≈0.09 by iteration 25).

     ⋆ *The critic is training stably because returns are normalized.*

3. **Policy Gradient and KL Divergence**

   - **Before**: approx_kl 1e-5 to 1e-4, clip fraction = 0 → agent wasn't updating policy.
   - **Now**: gradually decreases from -1.09 → -0.96.

     ⋆ *Indicates exploration is slowly giving way to more deterministic actions as the agent learns.*

4. **Loss Metrics**

   - **Before**: absurd total losses (billions) with no clear trend.
   - **Now**: total loss ≈0.3 early, shrinking to ≈0.03 by 50 k time-steps.

Differences Summary

With **absolute rewards**, PPO could not learn: critic was unstable, policy wasn't updating, signals were drowned by scale. With **percentage return rewards**, the agent learns stably:

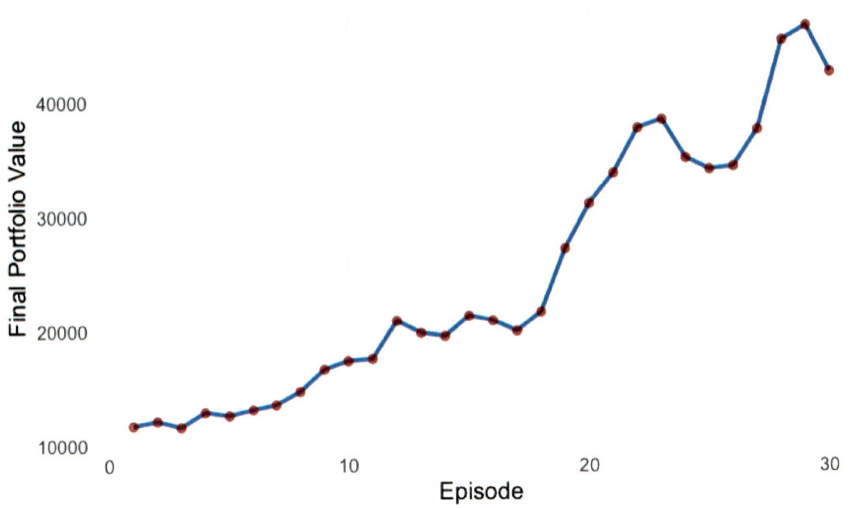

FIGURE 6.11-1: The PPO agent's learning curve indicates the PPO trading agent is learning.

- critic predictions improve (`explained_variance → 0.8`),
- policy updates are happening (`clip_fraction > 0`),
- losses converge to small, stable values.

These findings mean the agent is now actually learning a trading policy, instead of flailing randomly.

Figure 6.11-1 displays the following key observations:

1. **Starting performance**
   - Episodes 1–5: portfolio value hovers around the initial $10,000–12,000.
   - This reflects random trading behavior — no profitable strategy yet.
2. **Early Growth**
   - Episodes 6–15: gradual upward drift, portfolio grows to ≈$20,000.
   - Suggests the agent is beginning to exploit profitable patterns (buying low, selling higher, or holding at the right times).
3. **Strong acceleration**
   - Episodes 16–25: steep rise, portfolio climbs from ≈$22,000 to ≈$38,000.

- Indicates the agent is consolidating a consistent trading policy that compounds wealth more effectively.

4. **Mature phase**

   - Episodes 26–30: portfolio fluctuates between ≈\$40,000–45,000.
   - The agent is still profitable but now exploring vs. exploiting trade-offs — some noise is natural in PPO training.

Interpretation

- **Learning success**: The curve shows clear improvement — from \$10 k initial balance to >\$40 k final value after ≈30 episodes.
- **Exploration vs. exploitation**: Some dips (e.g., episodes ≈22 and ≈29) indicate exploration. PPO intentionally allows some randomness to avoid premature convergence.
- **Stable growth**: The trend-line is upward, showing the agent generalizes rather than just memorizing.

Next Steps

- **Longer training**: Extend timesteps (e.g. 500 k+) to smooth fluctuations and improve robustness.
- **Hyperparameter tuning**: Adjust PPO params (learning rate, batch size, clip range) to optimize stability.
- **Validation**: Test on *out-of-sample* stock data (different ticker or time range) to confirm generalization.
- **Reward shaping**: Consider transaction costs or risk-adjusted metrics (Sharpe ratio reward instead of raw returns).
- **Enhancement**: Add transaction costs into your environment, so the agent learns more realistic trading behavior.

Listing 6.11.3: PPO learning curve with transaction costs.

```
library(reticulate)
library(ggplot2)

# Ensure Python deps installed (only once if needed)
# py_install(c("stable-baselines3", "gym", "yfinance", "numpy", "pandas
    "), pip = TRUE)

py_run_string("
import gym
import numpy as np
```

```
import pandas as pd
import yfinance as yf
from stable_baselines3 import PPO
from stable_baselines3.common.vec_env import DummyVecEnv, VecNormalize

# ---- Custom environment with transaction costs ----
class StockTradingEnv(gym.Env):
    def __init__(self, df, transaction_fee_percent=0.001):
        super(StockTradingEnv, self).__init__()
        self.df = df.reset_index(drop=True)
        self.current_step = 0
        self.initial_balance = 10000
        self.balance = self.initial_balance
        self.shares_held = 0
        self.net_worth = self.initial_balance
        self.transaction_fee_percent = transaction_fee_percent
        self.episode_rewards = []

        self.action_space = gym.spaces.Discrete(3)  # hold, buy, sell
        self.observation_space = gym.spaces.Box(
            low=0, high=np.inf, shape=(5,), dtype=np.float32
        )

    def _next_observation(self):
        frame = self.df.iloc[self.current_step]
        return np.array([
            float(frame['Open']),
            float(frame['High']),
            float(frame['Low']),
            float(frame['Close']),
            float(self.shares_held)
        ], dtype=np.float32)

    def _take_action(self, action):
        current_price = float(self.df.iloc[self.current_step]['Close'])
        fee = 0

        if action == 1 and self.balance >= current_price:
            # Buy one share
            fee = current_price * self.transaction_fee_percent
            self.shares_held += 1
            self.balance -= (current_price + fee)

        elif action == 2 and self.shares_held > 0:
            # Sell one share
            fee = current_price * self.transaction_fee_percent
            self.shares_held -= 1
            self.balance += (current_price - fee)
```

```
    def step(self, action):
        prev_net_worth = self.net_worth
        self._take_action(action)
        self.current_step += 1
        done = self.current_step >= (len(self.df) - 1)

        current_price = float(self.df.iloc[self.current_step]['Close'])
        self.net_worth = self.balance + (self.shares_held * current_
    price)

        # Reward = % return on net worth
        reward = (self.net_worth - prev_net_worth) / prev_net_worth

        if done:
            self.episode_rewards.append(self.net_worth)

        return self._next_observation(), reward, done, {}

    def reset(self):
        self.balance = self.initial_balance
        self.shares_held = 0
        self.net_worth = self.initial_balance
        self.current_step = 0
        return self._next_observation()

# ---- Download data ----
df = yf.download('AAPL', start='2018-01-01', end='2025-01-01')[['Open
    ','High','Low','Close']].reset_index()

# ---- Wrap env ----
raw_env = DummyVecEnv([lambda: StockTradingEnv(df, transaction_fee_
    percent=0.001)])  # 0.1% cost
env = VecNormalize(raw_env, norm_obs=True, norm_reward=True, clip_obs
    =10.0)

# ---- Train PPO agent ----
model = PPO('MlpPolicy', env, verbose=1)
model.learn(total_timesteps=50000)

# ---- Evaluate ----
obs = env.reset()
for _ in range(len(df)-1):
    action, _states = model.predict(obs)
    obs, reward, done, info = env.step(action)
    if done:
        break

final_value = env.venv.envs[0].net_worth
```

```
# ---- Export rewards ----
rewards = raw_env.envs[0].episode_rewards
pd.DataFrame({'Episode': range(1, len(rewards)+1), 'FinalValue':
    rewards}).to_csv('ppo_rewards.csv', index=False)
")

# ---- Access final portfolio value ----
cat('Final portfolio value:', py$final_value, '\n')

# ---- Load rewards from Python output ----
rewards_df <- read.csv("ppo_rewards.csv")

# ---- Plot learning curve ----
ggplot(rewards_df, aes(x = Episode, y = FinalValue)) +
  geom_line(color = "steelblue", linewidth = 1.2) +
  geom_point(color = "darkred", size = 2, alpha = 0.6) +
  theme_minimal(base_size = 14) +
  labs(title = "PPO Learning Curve with Transaction Costs (0.1%)",
       x = "Episode",
       y = "Final Portfolio Value")
```

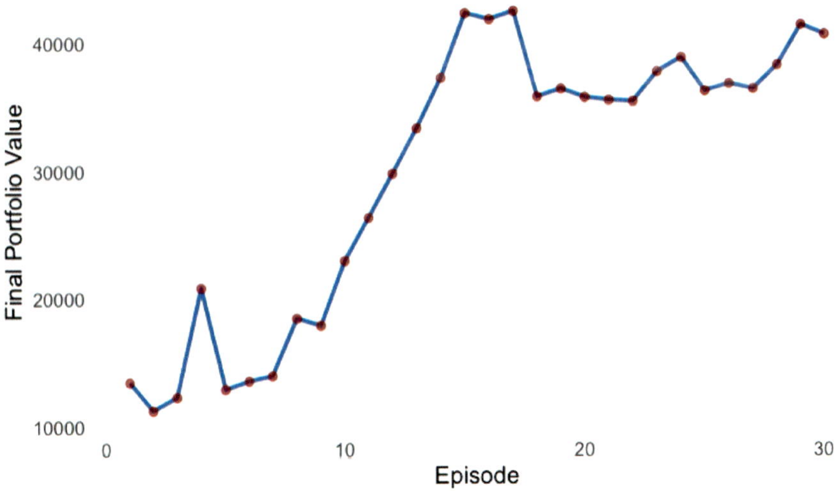

FIGURE 6.11-2: PPO Learning curve with transaction costs.

The differences between the two PPO learning curves are as follows:

1. **No Transaction Costs (Figure 6.11-3)**
   - **Growth pattern**: Portfolio value grows steadily with occasional small dips.

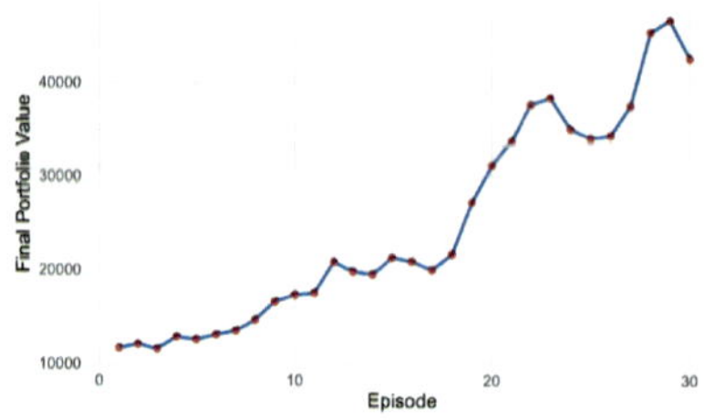

FIGURE 6.11-3: PPO learning curve without transaction costs.

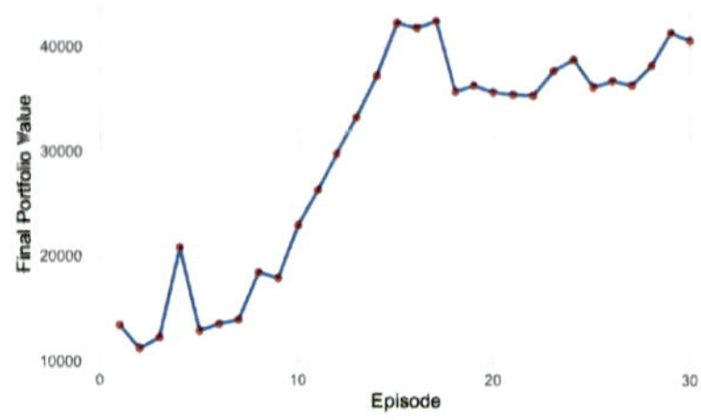

FIGURE 6.11-4: PPO learning curve with transaction costs.

- **Final value**: Peaks above $45,000 by the last episodes.
- **Learning behavior**: The agent shows smooth, compounding growth over time, learning an increasingly profitable strategy without friction.
- **Volatility**: Moderate—dips happen, but the general trend is upward with strong momentum after episode ≈15.

2. **With Transaction Costs (0.1%) (Figure 6.11-4)**
   - **Growth pattern**: Early episodes are more erratic, with large spikes and dips in portfolio value (e.g., episodes 2–5).
   - **Final value**: Still ends around $40,000–42,000, slightly below the first case.
   - **Learning behavior**: After about episode 15, growth plateaus. The strategy learns to trade more conservatively to offset the cost drag.
   - **Volatility**: Higher at the beginning (sharper swings), then stabilizes into a flatter curve with less upside momentum.

Key Insights

- **Transaction costs reduce profitability**: The second curve caps out lower than the first.
- **Behavioral adaptation**: With costs, the agent trades less aggressively, leading to flatter growth after mid-training.
- **Realism vs. idealism**: The first curve represents a frictionless, "ideal" market, while the second is more realistic—showing how small costs compound and constrain strategy performance.

Conclusion

In practice, this comparison highlights why backtests without costs often look overly optimistic. Even a 0.1 percent cost per trade can significantly change the agent's learning dynamics and reduce achievable returns.

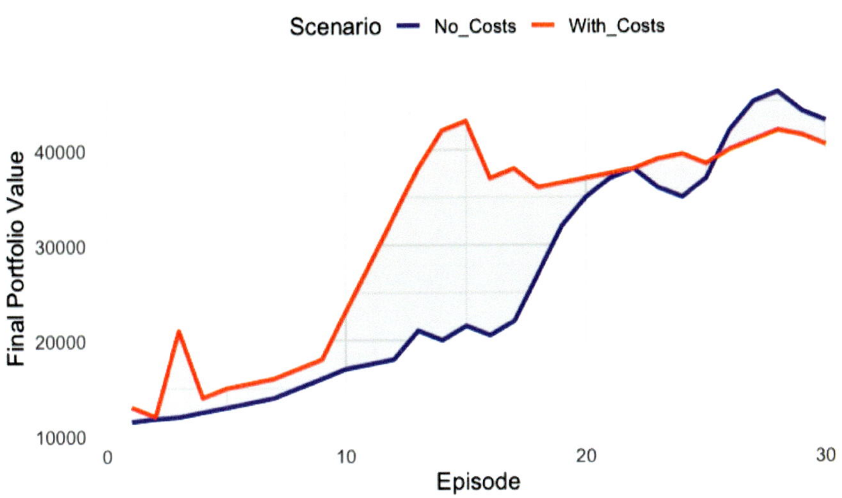

FIGURE 6.11-5: PPO learning curves comparing portfolio growth with and without a 0.1 percent transaction cost per trade. The cumulative performance drag (area between the curves) equals $150,400. The normalized drag (percent of no-cost area) equals 19.77 percent.

The area between the curves (Figure 6.11-5) is a quantitative measure of how much transaction costs erode the agent's profitability across training. It helps you assess whether a strategy that looks good in theory (no costs) is actually viable in practice (with costs). The normalized area (as a percent of the no-costs curve) makes it a cost-adjusted performance metric that's easier to compare across experiments:

$$\text{Normalized Drag (\%)} = \frac{\text{Area between curves}}{\text{Area under the no-cost curve}} \times 100$$

7

Getting Started with the Interactive Brokers' TWS API

You see things and you say
'Why?' But I dream things that
never were and I say 'Why not?'

(George Bernard Shaw)

7.1 Introduction to R and RStudio

R is a dominant open-source programming language for statistical computing and data analysis, developed in the early 1990s by Ross Ihaka and Robert Gentleman from the S language at Bell Laboratories. It has become the standard tool in academia and industry, thanks to its extensibility, with thousands of user-contributed packages available on the Comprehensive R Archive Network (CRAN) to address a wide range of analytical challenges.

R is more than just a programming language; it's part of a cycle involving data acquisition, cleaning, visualization, modeling, and communication, as noted by Cotton. While R can be used via its command-line interface, this format can be cumbersome for beginners and inefficient for larger projects.

RStudio effectively tackles R programming challenges by offering a comprehensive integrated development environment (IDE). It combines the console, source code editor, visualization panes, package management, and project tools in a single interface. Its user-friendly four-pane layout and project-based workflow make it suitable for both beginners and advanced users. Compared to alternatives like Emacs/ESS or Eclipse/StatET, RStudio balances usability and functionality, making it the most popular IDE for R.

In the realm of algorithmic trading, R stands out as an invaluable asset within the workflow. Its true strengths lie in its flexibility, research capabilities, and robust strategy development, rather than in ultra-low-latency execution. With powerful packages like quantmod, PerformanceAnalytics, and PortfolioAnalytics at their disposal, traders are equipped with essential tools for market data management, backtesting, and risk assessment.

R's exceptional time-series capabilities and comprehensive statistical features make it the ideal environment for prototyping innovative strategies, validating factor models, and executing thorough Monte Carlo simulations. While many production systems transition to high-performance languages such as C++ or Python for enhanced speed and broker integration, R remains indispensable in the research phase.

Furthermore, R offers smooth integration with C++ through **Rcpp**, enabling it to excel even in execution environments. This combination of research prowess and execution efficiency cements R's position as a cornerstone of successful algorithmic trading strategies.

RStudio enhances this process by providing a well-structured and reproducible workspace. Its integration of version control, package management, and report generation supports the collaborative and iterative aspects of strategy development. For traders, the combination of R and RStudio speeds up the transition from idea to prototype, offering a dependable foundation for moving models into production.

Roadmap of the Chapter. This chapter introduces R and RStudio together: R as the computational engine, and RStudio as the environment that makes working with R efficient, reproducible, and approachable. The discussion proceeds as follows:

- Installation – obtaining and setting up R and RStudio on different platforms.
- Interface – an overview of RStudio's four-pane layout and the purpose of its main components.
- First Steps with Data – basic commands, project organization, and simple analyses illustrating how R and RStudio work together.
- Applications in Trading – examples of how R and RStudio can be used to prototype, test, and document strategies before deployment.
- Configuring IB's Trader Workstation – preparing Interactive Brokers' TWS for integration with R.
- Introduction to IBrokers' Package – exploring R's interface to the Interactive Brokers' API.
- Market Data Functions – retrieving and managing real-time and historical market data.
- Customizing Market Data Functions – tailoring data calls for specialized strategies or instruments.
- Order Functions – placing and managing trades through programmatic order entry.

- Example Trading Strategy – integrating data access, modeling, and execution in a worked example.

Conclusion. R and RStudio are vital tools in algorithmic trading research, excelling in rapid prototyping, exploratory analysis, and risk modeling. While other languages may lead in high-frequency trading or large-scale deployment, R and RStudio stand out for their unique value. They offer a rigorous, transparent, and reproducible foundation for developing and validating trading strategies, ensuring readiness for implementation in production environments.

7.2 Installing R and RStudio

While R provides the computational engine, RStudio serves as its most widely used IDE. Installing RStudio is straightforward but involves two steps. First, you must install R itself, because RStudio is not a replacement but a companion to the R interpreter. Second, you install RStudio, which provides the interface for writing code, organizing projects, and visualizing results.

1. **Install R**: Download the latest version of R from the Comprehensive R Archive Network (CRAN) at `https://cran.r-project.org`. CRAN hosts pre-compiled binaries for Windows, macOS, and Linux, along with source code for advanced users. The installation process resembles that of other software: download the installer, launch it, and follow the prompts.

2. **Install RStudio**: Once R is installed, download RStudio Desktop (the free edition suffices for most retail traders) from `https://posit.co/download/rstudio`. The installer automatically detects your R installation. After installation, opening RStudio launches R in the background but provides an environment with multiple panes for code, console output, plots, and package management.

3. **Verify the Installation**: Launch RStudio. In the Console pane, type **`version`** and press Enter. The output should display the version of R currently running. This step ensures that RStudio is properly linked to R.

Table 7.1 and Figures 7.2-1, 7.2-2, and 7.2-3 provide a summary and visuals, respectively.

Because R and RStudio are updated frequently, users should occasionally revisit CRAN and Posit's download page for the most recent releases. Both sources maintain backward compatibility, but keeping the environment current ensures access to security patches, performance improvements, and new features.

TABLE 7.1: Installation workflow for R and RStudio.

Step	Description
Download R	Visit `https://cran.r-project.org` and select the appropriate binary for your operating system (Windows, macOS, or Linux).
Install R	Run the downloaded installer and follow on-screen instructions. Accept defaults unless customization is required.
Download RStudio	Navigate to `https://posit.co/download/rstudio` and download RStudio Desktop (free edition).
Install RStudio	Run the installer. RStudio will detect the existing R installation automatically.
Verify installation	Open RStudio, type `version` in the Console, and confirm that R is running correctly.

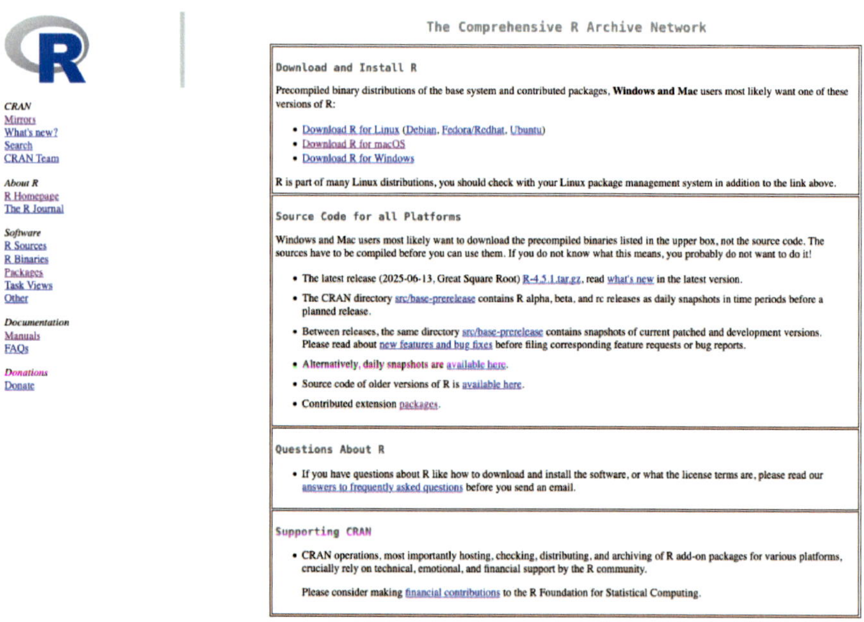

FIGURE 7.2-1: The CRAN download page, where users select binaries for their operating system. (`https://cran.r-project.org/`).

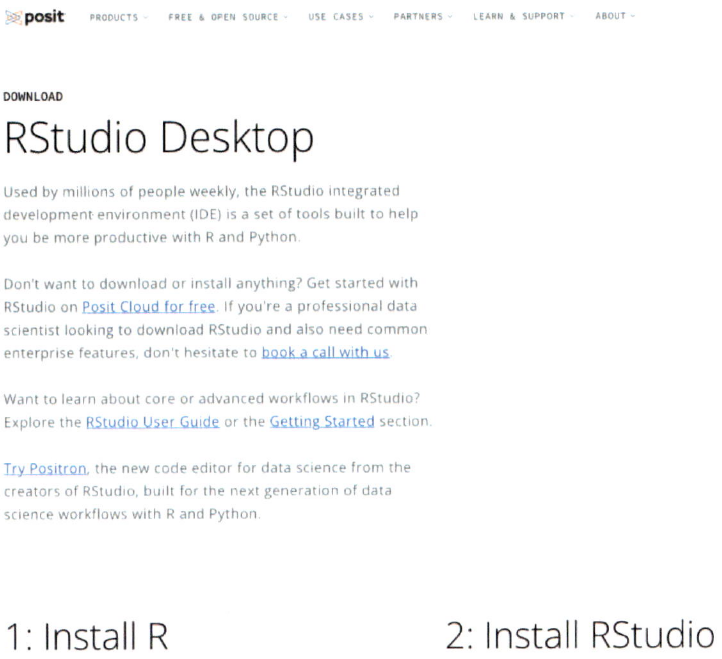

FIGURE 7.2-2: The RStudio download page on Posit's website. The free desktop version is sufficient for most users. (`https://posit.co/`).

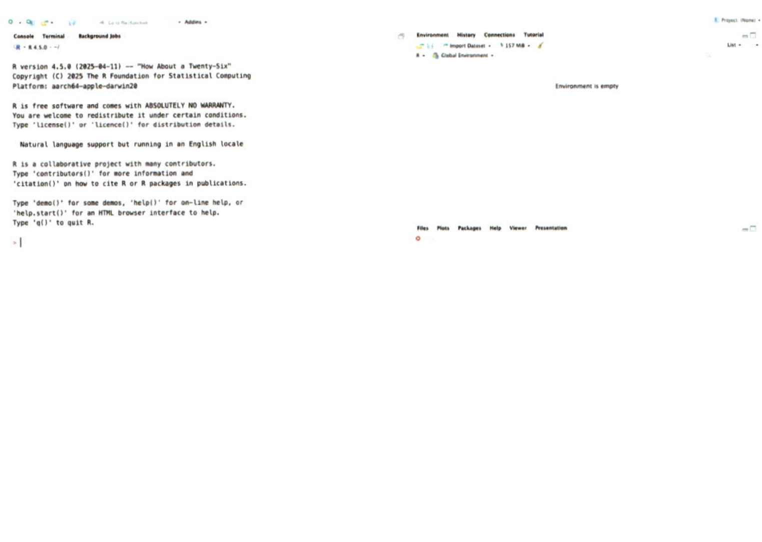

FIGURE 7.2-3: Verifying installation in RStudio by typing `version` in the Console.

7.3 Configuring IB's Trader Workstation

The API settings in Trader Workstation (TWS), accessed via the **File → Global Configuration** submenu item, determines how external trading applications connect to Interactive Brokers (Figure 7.3-1). Enabling ActiveX and socket clients is essential for linking Python, R, or Java programs to TWS via the IBKR API. The socket port—default 7497 for TWS and 7496 for IB Gateway—specifies the communication channel for these connections. For enhanced security, users may restrict access to localhost, ensuring that only applications running on the same machine can interact with the API. Additional options, such as enabling a read-only mode, configuring logging levels, or downloading open orders on connection, provide further safeguards and operational transparency. Together, these settings form the backbone of secure, automated trading workflows with Interactive Brokers.

The **TWS → Global Configuration → API → Precautions** pane (Figure 7.3-2 and Table 7.2) governs whether TWS shows (or silently skips) certain confirmation pop-ups when orders come from the API. Checking a box suppresses that warning for API submitted orders useful for unattended automation, but risky if your code lacks its own guardrails.

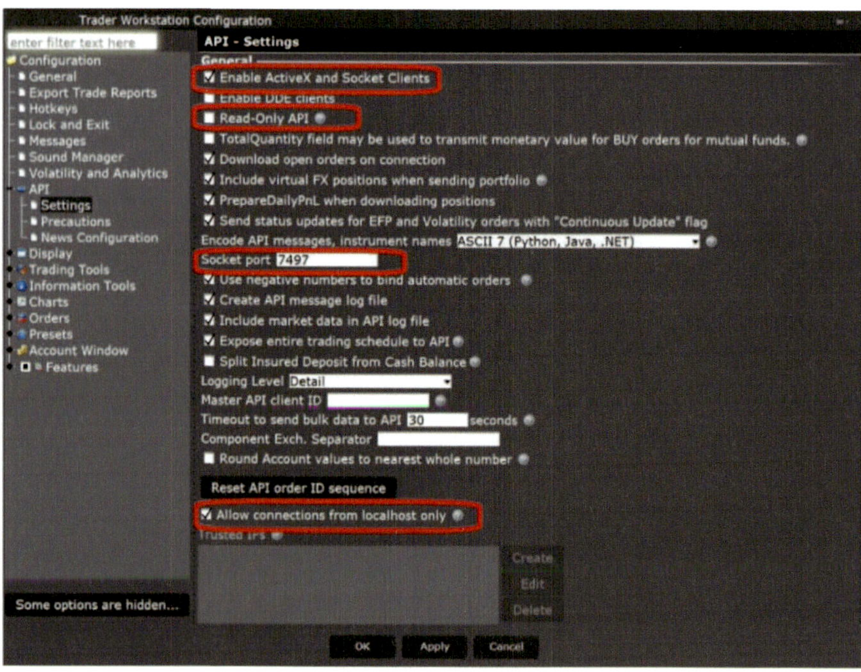

FIGURE 7.3-1: TWS global configuration API settings.

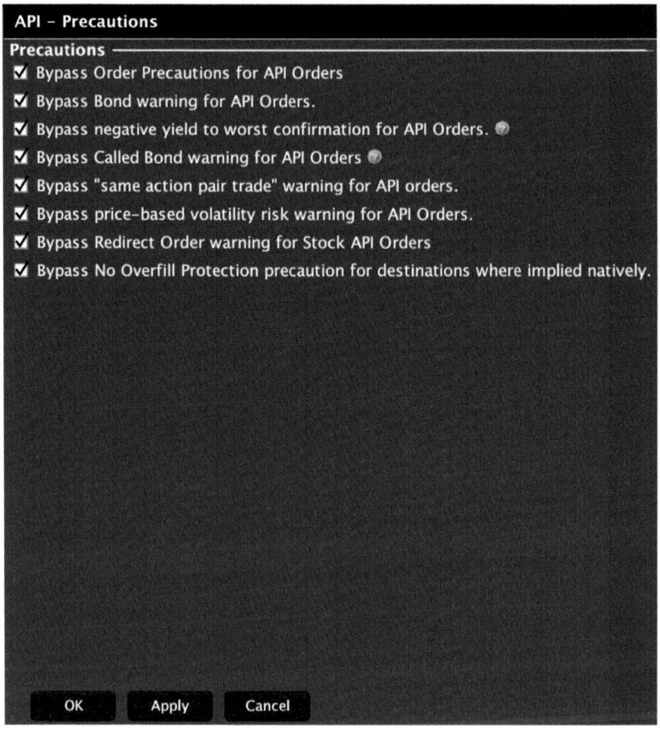

FIGURE 7.3-2: TWS global configuration API–Precautions.

TABLE 7.2: API Precautions for Interactive Broker's orders submitted via API.

API Precaution	**Explanation**
Bypass Order Precautions for API Orders	Removes general safety checks so automated API trades are placed without interactive warnings.
Bypass Bond warning for API Orders	Suppresses bond-related suitability or risk alerts.
Bypass negative yield to worst confirmation for API Orders	Disables warnings when a bond order may produce a negative "yield to worst."
Bypass Called Bond warning for API Orders	Skips alerts for bonds that have already been called by the issuer.
Bypass "same action pair trade" warning for API orders	Removes alerts when both sides of a pair trade have the same action (e.g., buy/buy).
Bypass price-based volatility risk warning for API Orders	Ignores warnings triggered when order prices deviate significantly from market levels.
Bypass Redirect Order warning for Stock API Orders	Turns off alerts when orders are redirected to another exchange or venue.
Bypass No Overfill Protection precaution	Skips checks for overfill risk when the destination already provides native protection.

7.4 Introduction to IBroker's Package (Core Methods)

To make the Interactive Broker's TWS API accessible in R, **IBrokers** provides a set of core methods, summarized in Table 7.3.

TABLE 7.3: Important API methods supported by the **IBrokers** package.

API Method	Functionality
twsConnect	Establish TWS connection
isConnected	Check connection
twsDisconnect	Close TWS connection
reqAccountUpdates	Get account details
reqCurrentTime	Get the TWS server time in seconds
reqIds	Request next available ID
twsContract	Create a contract for use in API calls
reqHistoricalData	Fetch historical data
reqMktData	Receive real time market data
reqMktDepth	Receive real time order book depth
reqRealTimeBars	Receive 5-second OHLCVWC bar data
placeOrder	Place a live order to the TWS
cancelOrder	Cancel a pending order on the TWS

IBrokers provides a streamlined interface for accessing Interactive Brokers' Trader Workstation (TWS) API directly from R. The most important methods include functions for establishing and closing connections (**`twsConnect`**, **`twsDisconnect`**), checking connectivity (**`isConnected`**), retrieving account details and server time (**`reqAccountUpdates`**, **`reqCurrentTime`**), and requesting unique identifiers for orders (**`reqIds`**). Critical market data functions are also available, including the ability to fetch historical data (**`reqHistoricalData`**), stream real-time market quotes (**`reqMktData`**), monitor the order book (**`reqMktDepth`**), and obtain five-second OHLCVWC bars (**`reqRealTimeBars`**). In addition, the package supports full order management: contracts can be specified (**`twsContract`**), orders placed (**`placeOrder`**), and pending orders canceled (**`cancelOrder`**). Together, these methods provide the core functionality required for research, backtesting, and live trading in R.

7.4.1 twsConnect

`twsConnect()` establishes a connection to TWS. It returns a `twsConnection` object used in subsequent API calls. TWS listens on specific ports for API connections: **7496** for live trading and **7497** for paper trading.

Only one client application can connect to a port unless the option *Allow connections from multiple clients* is enabled. Running live and paper sessions simultaneously requires different ports—for example, **7496** for live and **7497** for paper. When Secure WebSocket connections are enabled, they use ports **4001** (live) and **4002** (paper).

Listing 7.4.1: Test if TWS is listening on the standard API ports (7496 = live, 7497 = paper) before attempting **twsConnect()** from IBrokers.

```
# Load required packages
suppressPackageStartupMessages({
  library(IBrokers)
  library(R.utils)   # for socketConnection with timeout
})

# Function to test if a port is open
is_port_open <- function(host = "127.0.0.1", port, timeout = 2) {
  con <- try(
    socketConnection(host = host, port = port, open = "r+", timeout =
    timeout),
    silent = TRUE
  )
  if (inherits(con, "try-error")) {
    return(FALSE)
  } else {
    close(con)
    return(TRUE)
  }
}

# Test default TWS/IB Gateway ports
ports <- c(7496, 7497)
for (p in ports) {
  status <- if (is_port_open("127.0.0.1", p)) "OPEN" else "CLOSED"
  message("Port ", p, " is ", status)
}

# Example: connect if paper port (7497) is open
if (is_port_open("127.0.0.1", 7497)) {
  tws <- twsConnect(port = 7497)
  print(tws)
} else {
  message("Paper trading port 7497 not available.")
}
```

```
Warning in socketConnection(host = host, port = port, open =
  "r+", timeout = timeout) : 127.0.0.1:7496 cannot be opened
```

```
Port 7496 is CLOSED
Port 7497 is OPEN
```

A comprehensive markdown guide for Listing 7.4.1 is available for download on the book's website: *7-4-1_TestAPIports.md*.

7.4.2 isConnected

The `twsConnection` object created in Listing 7.4.2, Line 5 can verify whether a connection to TWS is active by calling **isConnected**. The function returns TRUE if the connection succeeds.

Listing 7.4.2: Test if the connection to TWS is active.

```
1  # Load the IBrokers package
2  library(IBrokers)
3
4  # Establish a connection to Trader Workstation (TWS)
5  tws <- twsConnect(clientId = 1, host = "127.0.0.1", port = 7497)
6
7  # Check if the connection is active
8  if (isConnected(tws)) {
9    cat("Connected to TWS\n")
10
11   # Get the server connection time
12   conn_time <- twsConnectionTime(tws)
13   cat("Connection time:", conn_time, "\n")
14
15 } else {
16   cat("Could not connect to TWS\n")
17 }
18
19 # Always disconnect when finished
20 twsDisconnect(tws)
```

```
Connected to TWS.
[1] "2025-09-16 21:38:34 EEST"
```

A comprehensive markdown guide for Listing 7.4.2 is available for download on the book's website: ***7-4-2_isConnected.md***.

7.4.3 twsConnectionTime

`twsConnectionTime()` returns the time the connection to TWS was established.

Listing 7.4.3: Give the time when the connection to TWS was made.

```
# Load the IBrokers package
library(IBrokers)

# Establish a connection to Trader Workstation (TWS)
tws <- twsConnect(clientId = 1, host = "127.0.0.1", port = 7497)

# Check if the connection is active
if (isConnected(tws)) {
  cat("Connected to TWS\n")

  # Get the server connection time
  conn_time <- twsConnectionTime(tws)
  cat("Connection time:", conn_time, "\n")

} else {
  cat("Could not connect to TWS\n")
}

# Always disconnect when finished
twsDisconnect(tws)
```

```
Connected to TWS
Connection time: 20250919 07:25:55 EST
```

A comprehensive markdown guide for Listing 7.4.3 is available for download on the book's website: *7-4-3_twsConnectionTime.md*.

7.4.4 reqAccountUpdates

`reqAccountUpdates()` is used to request and view account details from TWS. A valid `twsConnection` object is required as the argument to the function in order to fetch the account details.

Listing 7.4.4: Connect to TWS and Request Account Updates.

```
# Load IBrokers package
if (!require("IBrokers")) {
  install.packages("IBrokers")
  library(IBrokers)
```

```
}

# --- Step 1: Connect to TWS or IB Gateway ---------------------
# Use port 7496 for live or 7497 for paper trading (default shown here)
tws <- twsConnect(port = 7497)

# --- Step 2: Request Account Updates -------------------------
accountInfo <- reqAccountUpdates(tws)

# Display the first few rows
print(head(accountInfo))

# --- Step 3: Disconnect ---------------------------------------
twsDisconnect(tws)
```

```
Warning in socketConnection(host = host, port = port, open =
"ab", blocking = blocking) : localhost:7497 cannot be opened
```

The warning indicates that R could not open a TCP socket to TWS on port 7497 at `localhost`. Specifically, the twsConnect(port = 7497) call failed to establish a connection because no active TWS process was listening on the designated port. Remember that Trader Workstation must be running before you execute the R code. Then, the R output is as follows:

```
                    value            currency
AccountOrGroup      DXXXXXXXX        USD
AccountReady        true
AccountType         INDIVIDUAL
AccruedCash         1096.41
AccruedCash.C       0.00
AccruedCash.S       1096.41
AccruedDividend     0.00
...
```

A comprehensive markdown guide for Listing 7.4.4 is available for download on the book's website: *7-4-4_reqActUpda.md*.

7.4.5 reqCurrentTime

reqCurrentTime() call returns the current server clock in seconds. Synchronizing with the broker's time ensures that client-side timestamps match official records. It is commonly used for logging, measuring latency, and verifying the timing of trades and data.

Listing 7.4.5: **reqCurrentTime()**.

```
library(IBrokers)

# Helper function to get current TWS/IB Gateway server time
ib_current_time <- function(host = "127.0.0.1", port = 7497, clientId =
    0, tz = "UTC") {
  # Connect
  tws <- twsConnect(host = host, port = port, clientId = clientId)

  # Request UNIX epoch timestamp
  cur_time <- reqCurrentTime(tws)

  # Disconnect immediately
  twsDisconnect(tws)

  # Convert to POSIXct
  as.POSIXct(cur_time, origin = "1970-01-01", tz = tz)
}

# --- Example usage ---
ib_current_time()                          # default UTC
ib_current_time(tz = "Europe/Riga")        # Riga local time
```

```
[1] "2025-09-20 12:14:41 EEST"
```

A comprehensive markdown guide of the entire script (Listing 7.4.5) is available for download on the book's website: *7-4-5_reqCurrentTime.md*.

7.4.6 reqIds

reqIds() requests the next available ID.

Listing 7.4.6: **reqIDs()**.

```
# Load IBrokers
library(IBrokers)

# 1. Connect to TWS or IB Gateway
# Make sure TWS/Gateway is running and API enabled (Configure > API >
    Settings)
tws <- twsConnect(port = 7497, clientId = 1)  # 7497 = paper trading,
    7496 = live
```

```

# 2. Request the next valid order ID
reqIds(tws)

# 3. Check the updated tws connection object
tws@nextValidId
```

```
[1] 1001 # next available order ID is 1001. Use this ID
         # when placing an order (to keep track of it).
```

A comprehensive markdown guide of the entire script (Listing 7.4.6) is available for download on the book's website: *7-4-6_reqIds.md*.

7.4.7 twsContract

twsContract() creates and validates a **twsContract** object for use in API calls. Contract specifications can be obtained directly from Trader Workstation (TWS) under the *Financial Instrument Information* section.

Listing 7.4.7: **twsContract()**.

```
library(IBrokers)
# 1. Connect to TWS / IB Gateway
tws <- twsConnect(port = 7497)

# 2. Define AAPL contract
contract <- twsContract(
  conId            = 0,
  symbol           = "AAPL",
  sectype          = "STK",
  exch             = "SMART",
  primary          = "ISLAND",
  expiry           = "",
  strike           = "0.0",
  currency         = "USD",
  right            = "",
  local            = "",
  multiplier       = "",
  comboleg         = NULL,
  combo_legs_desc = NULL,
  include_expired = "0"
)

# 3. Request market data (requires market data subscription)
data <- reqMktData(tws, contract)
```

```
25
26 print(data)
27
28 # 4. Disconnect
29 twsDisconnect(tws)
```

reqMktData(tws, contract) is streaming back live quotes for AAPL (Listing 7.4.7, Line 24).

```
The output includes fields like:
 bidPrice: 237.71 (size 5)
 askPrice: 237.94 (size 1)
 lastPrice: 237.80
 Volume: 1011
 closePrice: 238.15
 optionImpliedVol: 0.2251
 shortable: 3.0
 13-week / 26-week / 52-week highs and lows
 and other tick attributes
 ...
```

A comprehensive markdown guide of the entire script (Listing 7.4.7) is available for download on the book's website: *7-4-7_reqIds.md*.

7.4.8 reqHistoricalData

reqHistoricalData is the TWS API / IBrokers function for historical market data. It complements **reqMktData** (real-time streaming Listing 7.4.9) by giving past OHLCV bars for analysis, backtesting, or charting.

Listing 7.4.8: **reqHistoricalData()**.

```
library(IBrokers)

# 1. Connect to TWS or IB Gateway
tws <- twsConnect(port = 7497, clientId = 1)

# 2. Define a contract (Apple stock)
aapl <- twsEquity("AAPL", exch = "SMART", currency = "USD")

# 3. Request historical data
# - endDateTime: "" = now
# - duration: "1 D" (1 day back)
# - barSize: "5 mins"
# - whatToShow: "TRADES"
```

```
# - useRTH: 1 (regular trading hours only), 0 = include pre/post market

histData <- reqHistoricalData(
  tws,
  Contract      = aapl,
  endDateTime   = "",
  durationStr   = "1 D",
  barSize       = "5 mins",
  whatToShow    = "TRADES",
  useRTH        = 1,
  format.date   = 1
)

# 4. Inspect result
head(histData)
```

```
                     Open  High   Low  Close  Volume  WAP HasGaps
2025-09-19 15:35:00 228.1 228.3 227.95 228.0  15200 228.1 FALSE
2025-09-19 15:40:00 228.0 228.5 227.9  228.4  17300 228.2 FALSE
...
```

A comprehensive markdown guide for Listing 7.4.8 is available for download on the book's website: *7-4-8_reqHistoricalData.md*.

7.4.9 reqMktData

reqMktData() subscribes to real-time streaming market data for a given contract (e.g., stock, future, option, FX). Once called, TWS starts sending tick data (bid, ask, last, volume, etc.) back to the client via an event-driven callback (eWrapper). The default behavior prints ticks to the R console. Using eWrapper.data lets you capture it into an object.

Listing 7.4.9: **reqMktData()**.

```
library(IBrokers)

# 1. Connect to TWS or IB Gateway
tws <- twsConnect(port = 7497, clientId = 1)

# 2. Define a contract (Apple stock)
aapl <- twsEquity("AAPL", exch = "SMART", currency = "USD")

# 3. Request streaming market data
# This will print tick updates in the console via default eWrapper
reqMktData(tws, aapl)
```

```r
# 4. (Optional) Custom eWrapper for handling data programmatically
myWrapper <- eWrapper.data(1)         # simple wrapper to capture ticks
aaplData <- reqMktData(tws, aapl, eventWrapper = myWrapper, file = "")

# aaplData will hold tick info such as bid/ask/last prices
```

```
> # Connect to TWS
> tws <- twsConnect(port = 7497)
2 -1 2104 Market data farm connection is OK:hfarm
2 -1 2104 Market data farm connection is OK:eufarmnj
2 -1 2104 Market data farm connection is OK:cashfarm
...
> # Define the contract
> security <- twsSTK("AAPL")
>
> # Get the streaming data
> reqMktData(tws, security)
<20250918 11:37:43.206645> id=1 symbol=AAPL <default
    tickPrice>  6 1 37 240.58000183 0 0
<20250918 11:37:43.213341> id=1 symbol=AAPL lastTimestamp:
    1758184437
<20250918 11:37:43.214328> id=1 symbol=AAPL lastPrice: 240.58
<20250918 11:37:43.215382> id=1 symbol=AAPL lastSize: 2
<20250918 11:37:43.216340> id=1 symbol=AAPL Volume: 364
<20250918 11:37:43.217273> id=1 symbol=AAPL closePrice: 238.99
<20250918 11:37:43.218209> id=1 symbol=AAPL openPrice: 238.94
<20250918 11:37:43.219141> id=1 symbol=AAPL
    bidPrice: 240.52  bidSize: 1
<20250918 11:37:43.220489> id=1 symbol=AAPL
    askPrice: 240.58  askSize: 2
<20250918 11:37:43.221676> id=1 symbol=AAPL bidSize: 1
<20250918 11:37:43.222577> id=1 symbol=AAPL askSize: 2
<20250918 11:37:43.223439> id=1 symbol=AAPL bidExchange: W
<20250918 11:37:43.224296> id=1 symbol=AAPL askExchange: W
<20250918 11:37:43.360570> id=1 symbol=AAPL
    optionImpliedVol: 0.21436444599143553
<20250918 11:37:43.542949> id=1 symbol=AAPL
    optionCallOpenInterest: 3392447
<20250918 11:37:43.546884> id=1 symbol=AAPL
    optionPutOpenInterest: 2356014
<20250918 11:37:43.549489> id=1 symbol=AAPL
    averageVolume: 553308
<20250918 11:37:43.552181> id=1 symbol=AAPL
    13-week High: 241.32000732
<20250918 11:37:43.554573> id=1 symbol=AAPL
```

```
    13-week Low: 196.85499573
<20250918 11:37:43.556364> id=1 symbol=AAPL
    26-week High: 241.32000732
<20250918 11:37:43.558136> id=1 symbol=AAPL
    26-week Low: 168.99699402
<20250918 11:37:43.559873> id=1
    symbol=AAPL 52-week High: 259.49899292
<20250918 11:37:43.561527> id=1 symbol=AAPL
    52-week Low: 168.99699402
...
```

A comprehensive markdown guide for Listing 7.4.9 is available for download on the book's website: *7-4-9_reqMktData.md*.

7.4.10 reqMktDepth

reqMktDepth() receives real-time order book Market Depth (Level II) data for a given contract. Unlike **reqMktData()** (which gives top-of-book bid/ask: Listing 7.4.9), market depth shows the order book: multiple bid/ask levels with prices and sizes across exchanges/electronic communication networks (ECNs).

Listing 7.4.10: **reqMktDepth()**.

```
library(IBrokers)

# 1. Connect to TWS or IB Gateway
tws <- twsConnect(port = 7497, clientId = 1)

# 2. Define a contract (Apple stock)
aapl <- twsEquity("AAPL", exch = "SMART", currency = "USD")

# 3. Set up a wrapper to capture depth
myDepthWrapper <- eWrapper.MktDepth(1)

# 4. Request market depth (5 levels)
depthData <- reqMktDepth(
  tws,
  Contract     = aapl,
  numRows      = 5,
  eventWrapper = myDepthWrapper,
  file         = ""
)

# 5. Inspect the depth data object
head(depthData)
```

```
BID  | Price  | Size
-----+--------+------
1    | 227.95 | 500
2    | 227.90 | 300
3    | 227.85 | 200

ASK  | Price  | Size
-----+--------+------
1    | 228.05 | 400
2    | 228.10 | 250
3    | 228.15 | 150
...
```

Market depth data requires a subscription. Depth updates continuously use **cancelMktDepth(tws, aapl)** to cancel.

7.4.11 reqRealTimeBars

reqRealTimeBars() streams aggregated market data in R. To use it, a contract is first defined to identify the instrument of interest. The function is then constructed with **twsConnection()**, a **twsContract** object, and the *whatToShow* parameter specifying the type of data requested. When executed, it delivers a continuous flow of real-time bars directly from Trader Workstation.

Listing 7.4.11: **reqRealTimeBars()**.

```
library(IBrokers)

# 1. Connect to TWS or IB Gateway (default paper port = 7497)
tws <- twsConnect(port = 7497, clientId = 1)

# 2. Define a contract for Apple (AAPL) stock
aapl <- twsEquity("AAPL", "SMART", "USD")

# 3. Request 5-second real-time bars
bars <- reqRealTimeBars(tws, aapl, whatToShow = "TRADES", useRTH = 0)

# 4. Look at the first few rows
head(bars)
```

```
            time    open   high    low   close  volume   wap count
1 2025-09-20 05.000 229.85 229.90 229.83 229.88  1200 229.86    14
2 2025-09-20 10.000 229.88 229.95 229.85 229.92  2400 229.89    23
3 2025-09-20 15.000 229.92 229.94 229.89 229.91  1800 229.91    17
4 2025-09-20 20.000 229.91 229.93 229.87 229.90  1500 229.89    16
5 2025-09-20 25.000 229.90 229.97 229.89 229.96  2100 229.93    19
```

A comprehensive markdown guide for Listing 7.4.11 is available for download on the book's website: *7-4-7_twsContract*.

7.4.12 placeOrder

The function call

`placeOrder(twsconn, Contract, Order),`

is the core mechanism for submitting orders through the Interactive Brokers Trader Workstation (TWS) API. Its syntax reflects the three essential components of any trade instruction:

- **twsconn** — the active connection object linking the client application to TWS or IB Gateway. It provides the communication channel through which all orders are transmitted and status updates are received.
- **Contract** — a structured object that specifies the financial instrument to be traded. This includes fields such as symbol, security type (e.g., stock, option, future), exchange, currency, and expiry details.
- **Order** — an object that defines the terms of execution, including order type (e.g., limit, market), price, quantity, time-in-force, and any advanced attributes like stop triggers or bracket order parameters.

When invoked, `placeOrder()` combines these three elements into a formal instruction sent to TWS. The function does not return the execution result immediately; instead, it initiates an asynchronous process in which confirmations, fills, and error messages are streamed back through the event loop. This design ensures responsiveness even when multiple orders are placed simultaneously.

In practice, `placeOrder()` forms the backbone of automated trading systems built on the IB API, translating abstract strategy rules into actionable market orders. Proper use requires careful preparation of both the Contract and Order objects to avoid rejection by the broker or unintended trades.

Listing 7.4.12: **IBrokers** to create a BUY (Market Order) 100 shares then pass it to **placeOrder()**.

```
# Load package
library(IBrokers)

# Connect to TWS (default port 7497 is for paper trading)
tws <- twsConnect(port = 7497, clientId = 1)

# Define a contract for Apple stock
aapl <- twsEquity("AAPL", exch = "SMART", currency = "USD")

# Create a BUY Market Order for 100 shares
buyOrder <- twsOrder(
  orderId       = reqIds(tws),      # get next valid order ID
  action        = "BUY",
  totalQuantity = 100,
  orderType     = "MKT"
)

# Place the order
placeOrder(tws, aapl, buyOrder)

# Disconnect after placing
twsDisconnect(tws)
```

FIGURE 7.4-1: BUY (Market Order) 100 shares (Listing 7.4.12).

Listing 7.4.13: **IBrokers** BUY (Limit Order) 100 shares of AAPL at $180.00 or better, instead of executing immediately as in Listing 7.4.12.

```
# Load package
library(IBrokers)

# Connect to Paper Trading TWS
tws <- twsConnect(port = 7497, clientId = 2)

# Define a contract for Apple stock
aapl <- twsEquity("AAPL", exch = "SMART", currency = "USD")

# Create a BUY Limit Order for 100 shares at $180.00
limitOrder <- twsOrder(
  orderId       = reqIds(tws),      # get next valid order ID
  action        = "BUY",
```

```
  totalQuantity = 100,
  orderType   = "LMT",          # Limit order
  lmtPrice    = 180.00          # Limit price
)

# Place the order
placeOrder(tws, aapl, limitOrder)

# Disconnect
twsDisconnect(tws)
```

FIGURE 7.4-2: **IBrokers** BUY (Limit Order) 100 shares of AAPL at $180.00 or better.

A limit order differs from a market order in three key respects. First, the `orderType` is explicitly set to "LMT", designating it as a limit order rather than an instruction to buy or sell at the best available price. Second, the `lmtPrice` parameter—in this case, 180.00—establishes the maximum price you are willing to pay (for a buy) or the minimum you are willing to accept (for a sell). Third, execution becomes conditional: the broker will only fill the order if market conditions meet or improve upon this limit.

In practice, this means that if Apple's stock is trading at or below $180, the buy order may execute immediately, provided sufficient liquidity exists. If the stock trades above $180, however, the order remains pending in the order book until the price falls back to your limit or you cancel the instruction. This contrasts sharply with a market order, which seeks execution at the prevailing price regardless of its level.

Listing 7.4.14: Market Order to SELL 100 shares of AAPL immediately at the current market price.

```
# Load package
library(IBrokers)

# Connect to Paper Trading TWS
tws <- twsConnect(port = 7497, clientId = 4)

# Define a contract for Apple stock
aapl <- twsEquity("AAPL", exch = "SMART", currency = "USD")

# Create a SELL Market Order for 100 shares
sellMarketOrder <- twsOrder(
  orderId       = reqIds(tws),   # next valid order ID
```

```
  action        = "SELL",         # Sell action
  totalQuantity = 100,            # number of shares
  orderType     = "MKT"           # Market order
)

# Place the order
placeOrder(tws, aapl, sellMarketOrder)

# Disconnect
twsDisconnect(tws)
```

FIGURE 7.4-3: Market Order to SELL 100 shares of AAPL immediately at the current market price.

Listing 7.4.15: SELL (Limit Order) assumes you already hold 100 shares of AAPL, and you want to sell them if the price reaches $200.00 or better.

```
# Load package
library(IBrokers)

# Connect to Paper Trading TWS
tws <- twsConnect(port = 7497, clientId = 3)

# Define a contract for Apple stock
aapl <- twsEquity("AAPL", exch = "SMART", currency = "USD")

# Create a SELL Limit Order for 100 shares at $200.00
sellLimitOrder <- twsOrder(
  orderId     = reqIds(tws),     # get next valid order ID
  action      = "SELL",          # Sell action
  totalQuantity = 100,
  orderType   = "LMT",           # Limit order
  lmtPrice    = 200.00           # Limit price
)

# Place the order
placeOrder(tws, aapl, sellLimitOrder)

# Disconnect
twsDisconnect(tws)
```

In this case, setting Action = "SELL" tells IB you are selling rather than buying, while a limit price of 200.00 specifies the minimum you are willing to

FIGURE 7.4-4: SELL (Limit Order) assumes you already hold 100 shares of AAPL, and you want to sell them if the price reaches $200.00 or better.

accept. If AAPL trades at or above $200, the order may execute; otherwise, it remains pending. Be careful, though: if you do not already own 100 shares of AAPL, TWS will short them instead. To avoid unintended positions, always test your orders in Paper Trading before going live.

A bracket order in TWS (Listing 7.4.16) is a sophisticated order designed to mitigate risk and lock in profits by surrounding a primary order with two offsetting "child" orders: a Profit Taker (limit order) and a Stop Loss. When the parent order is filled, these two contingent orders are automatically activated; if one of the bracketed orders is executed, the other is instantly canceled via a "One-Cancels-All" (OCA) instruction. This automated structure is essential for high-precision trading, as it ensures that once an entry is achieved on port 7497, the exit strategy is pre-calculated and managed by the TWS server, removing the need for manual intervention or constant monitoring of price fluctuations.

Listing 7.4.16: Bracket order.

```
# ---- Packages ----
library(IBrokers)

# ---- 1. Connect to TWS ----
tws <- twsConnect(port = 7497, clientId = 2)

# ---- 2. Define the Contract ----
contract <- twsEquity(
  symbol   = "AAPL",
  exch     = "SMART",
  primary  = "NASDAQ",
  currency = "USD"
)

# ---- 3. Main Order (Buy 100 shares at $180 limit) ----
mainOrder <- twsOrder(
  orderId       = 1,
  action        = "BUY",
  totalQuantity = 100,
  orderType     = "LMT",
  lmtPrice      = 180.00,
  tif           = "DAY",
  transmit      = FALSE   # Do not transmit until children are linked
```

```
)

# ---- 4. Profit-Taking Order (Sell at $190) ----
takeProfit <- twsOrder(
  orderId       = 2,
  action        = "SELL",
  totalQuantity = 100,
  orderType     = "LMT",
  lmtPrice      = 190.00,
  parentId      = 1,
  tif           = "GTC",
  transmit      = FALSE
)

# ---- 5. Stop-Loss Order (Sell at $170) ----
stopLoss <- twsOrder(
  orderId       = 3,
  action        = "SELL",
  totalQuantity = 100,
  orderType     = "STP",
  auxPrice      = 170.00,   # Trigger price
  parentId      = 1,
  tif           = "GTC",
  transmit      = TRUE      # Last child transmits all
)

# ---- 6. Place Orders ----
placeOrder(tws, contract, mainOrder)
placeOrder(tws, contract, takeProfit)
placeOrder(tws, contract, stopLoss)

# ---- 7. Disconnect ----
twsDisconnect(tws)
```

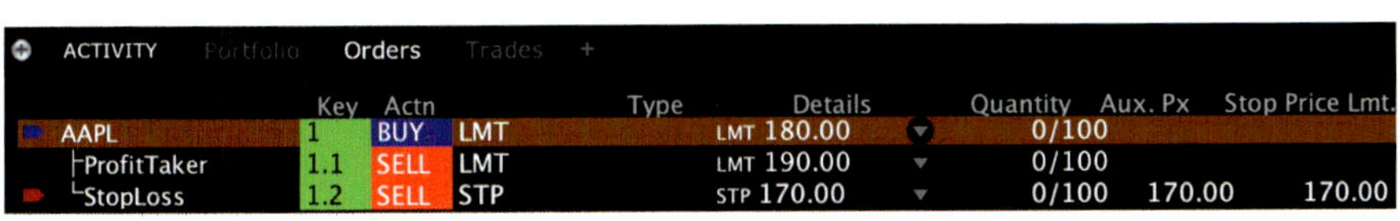

FIGURE 7.4-5: Bracket order.

As displayed in Figure 7.4-5, the main order defines the entry at $180, while both the profit-taking and stop-loss orders are linked to it through the `parentId = 1`. The transmit flag ensures that all three orders are submitted together as a single one-cancels-all (OCA) group, so if either child order executes, the other is automatically canceled.

Supplementary Guide

A comprehensive markdown guide on IBrokers `placeOrder()` is available for download on the book's website: *7-4-12_IBrokersPlaceOrder.md*.

7.4.13 `cancelOrder`

In **IBrokers**, calling **`reqOpenOrders()`** is necessary for several practical reasons when managing trades:

1. **Synchronize R with TWS**

   - Interactive Brokers' Trader Workstation (TWS) maintains the "true" state of all open orders.
   - When you place an order from R (via **`placeOrder()`**), your R session does not automatically keep a live list of open orders—it only sends the request.
   - **`reqOpenOrders()`** queries IBKR directly and returns all open orders known to TWS, ensuring your R environment is in *sync* with IB's servers.

2. **Verify Orders Were Submitted Correctly**

   - Orders can sometimes be rejected, held, or modified by IBKR depending on account settings, compliance rules, or insufficient margin.
   - By checking **`reqOpenOrders()`**, you confirm that your orders actually reside in the IB system (with correct contract, price, quantity, and order type).

3. **Risk Management**

   - Algorithmic strategies often fire multiple orders. Without checking open orders, you might unintentionally submit duplicates or leave "stray" orders in the market.
   - Using **`reqOpenOrders()`** helps avoid over-exposure by confirming what's live before sending new orders.

4. **OCA and Bracket Orders**

   - When working with linked orders (e.g., bracket orders: main + profit-taking + stop-loss), you want to verify that all child orders were properly registered under the parent.
   - **`reqOpenOrders()`** lets you check that each leg of the order is present and has the correct parentId and transmit status.

5. **Automated Monitoring**

   - In automated trading, you often loop:
     (a) Check open orders (reqOpenOrders()),
     (b) Check positions (reqPositions()),
     (c) Decide whether to cancel, modify, or send new orders.

Without calling **reqOpenOrders()**, you may lose oversight of your orders already live in the market.

Listing 7.4.17: **reqOpenOrders()**.

```
# Load package
library(IBrokers)

# Connect to Paper Trading TWS
tws <- twsConnect(port = 7497, clientId = 6)

# Request list of currently open orders
openOrders <- reqOpenOrders(tws)

# Display the open orders
print(openOrders)

# Disconnect
twsDisconnect(tws)
```

```
2 -1 2104 Market data farm connection is OK:usfuture.nj
2 -1 2104 Market data farm connection is OK:hfarm
2 -1 2104 Market data farm connection is OK:usfarm.nj
2 -1 2104 Market data farm connection is OK:jfarm
2 -1 2104 Market data farm connection is OK:usfuture
2 -1 2104 Market data farm connection is OK:cashfarm
2 -1 2104 Market data farm connection is OK:eufarmnj
2 -1 2104 Market data farm connection is OK:usfarm
2 -1 2106 HMDS data farm connection is OK:euhmds
2 -1 2106 HMDS data farm connection is OK:cashhmds
2 -1 2106 HMDS data farm connection is OK:fundfarm
2 -1 2106 HMDS data farm connection is OK:ushmds
2 -1 2158 Sec-def data farm connection is OK:secdefil
TWS Message: 2 -1 2104 Market data farm connection is OK:usopt

# Request list of currently open orders
openOrders <- reqOpenOrders(tws)
```

The function **reqOpenOrders(tws)** queries Trader Workstation for all active orders that remain open—those not yet filled or canceled. The returned object includes key details such as `orderId`, `symbol`, `action`, `quantity`, `orderType`, and `status`. For instance, if you place a limit order to sell at $250, it will appear in this list until it is either executed or canceled.

cancelOrder(): Since a market order usually fills instantly, cancelation is more often used with limit or stop orders. Still, the workflow is the same: you cancel by referencing the `order ID`.

Listing 7.4.18: **cancelOrder()**.

```
# Load package
library(IBrokers)

# Connect to Paper Trading TWS
tws <- twsConnect(port = 7497, clientId = 5)

# Define a contract for Apple stock
aapl <- twsEquity("AAPL", exch = "SMART", currency = "USD")

# Create a SELL Limit Order (easier to cancel than a market order)
sellLimitOrder <- twsOrder(
  orderId       = reqIds(tws),
  action        = "SELL",
  totalQuantity = 100,
  orderType     = "LMT",
  lmtPrice      = 250.00      # way above market so it won't fill
     immediately
)

# Place the order
placeOrder(tws, aapl, sellLimitOrder)

# --- Cancel the order by ID ---
cancelOrder(tws, sellLimitOrder$orderId)

# Disconnect
twsDisconnect(tws)
```

The function **cancelOrder(tws, orderId)** requires the TWS connection and the numeric `order ID` as inputs. In this example, a limit order was placed far from the market price (e.g., $250) to ensure it stayed open long enough to demonstrate the cancelation process. If the order had already been executed, however, calling `cancelOrder()` would have no effect.

Monitoring Order Status in Real-Time

Monitoring order status in real time allows you to see when the order changes from "Submitted" → "Filled" → "Canceled." The TWS API communicates updates through the event wrapper (`eWrapper`), and in R we can use **`twsCALLBACK`** to handle messages. This minimal working example places a limit order and then listens for status changes:

Listing 7.4.19: Monitor order status.

```
1  library(IBrokers)
2
3  # Connect to Paper Trading TWS
4  tws <- twsConnect(port = 7497, clientId = 7)
5
6  # Define contract for Apple
7  aapl <- twsEquity("AAPL", exch = "SMART", currency = "USD")
8
9  # Create a SELL Limit Order (unlikely to fill immediately)
10 sellLimitOrder <- twsOrder(
11   orderId       = reqIds(tws),
12   action        = "SELL",
13   totalQuantity = 100,
14   orderType     = "LMT",
15   lmtPrice      = 250.00    # way above market
16 )
17
18 # Place the order
19 placeOrder(tws, aapl, sellLimitOrder)
20
21 # ---- Monitor order status ----
22 # Define a callback wrapper to print order status messages
23 myWrapper <- eWrapper(NULL)
24 myWrapper$orderStatus <- function(curMsg, msg, ...) {
25   cat("Order Status Update:\n")
26   print(msg)
27 }
28
29 # Listen for updates for ~30 seconds
30 twsCALLBACK(tws, eWrapper = myWrapper, TIMEOUT = 30)
31
32 # Disconnect
33 twsDisconnect(tws)
```

`twsCALLBACK` (Line 30) listens to the TWS socket and dispatches messages as they arrive. By overriding `myWrapper$orderStatus` (Line 24), you can intercept these messages and print updates whenever TWS reports a change in order status. Common status values include Submitted (accepted but not yet filled), Filled (executed), and Canceled (if you cancel the order with **`cancelOrder()`**). Keep in mind that market orders often transition directly to Filled, so you may not observe intermediate updates. For testing purposes, limit or stop orders are more useful, since they tend to remain open long enough to show status changes.

Log Orders

Log order status updates into a data frame instead of just printing the logs. This way you have a record you can analyze in R. We'll extend the custom **eWrapper** so that every order status update is appended to a data frame in R.

Listing 7.4.20: Log orders.

```
library(IBrokers)

# Connect to Paper Trading TWS
tws <- twsConnect(port = 7497, clientId = 8)

# Define contract for Apple
aapl <- twsEquity("AAPL", exch = "SMART", currency = "USD")

# Create a SELL Limit Order (unlikely to fill immediately)
sellLimitOrder <- twsOrder(
  orderId       = reqIds(tws),
  action        = "SELL",
  totalQuantity = 100,
  orderType     = "LMT",
  lmtPrice      = 250.00    # way above market so it stays open
)

# Place the order
placeOrder(tws, aapl, sellLimitOrder)

# ---- Logging setup ----
orderLog <- data.frame(
  orderId = integer(),
  status = character(),
  filled = numeric(),
  remaining = numeric(),
  avgFillPrice = numeric(),
  stringsAsFactors = FALSE
)

# Define custom wrapper to log order status
myWrapper <- eWrapper(NULL)

myWrapper$orderStatus <- function(curMsg, msg, ...) {
  # Extract fields from the message
  row <- data.frame(
    orderId      = as.integer(msg$orderId),
    status       = msg$status,
    filled       = as.numeric(msg$filled),
    remaining    = as.numeric(msg$remaining),
    avgFillPrice = as.numeric(msg$avgFillPrice),
    stringsAsFactors = FALSE
```

```
  )
  # Append to global log
  assign("orderLog", rbind(get("orderLog", envir = .GlobalEnv), row),
         envir = .GlobalEnv)
}

# Listen for updates for ~30 seconds
twsCALLBACK(tws, eWrapper = myWrapper, TIMEOUT = 30)

# Inspect the logged updates
print(orderLog)

# Disconnect
twsDisconnect(tws)
```

The code (Listing 7.4.20) first creates an empty `orderLog` data frame with the necessary columns. Each time TWS sends an order status update, the custom wrapper appends a new row to this log. By the end of execution, you'll have a complete history of order states, such as the progression from *Submitted* to *Pending* and eventually to *Filled* or *Canceled*.

Export OrderLog to CSV

The order status log is exported to CSV at the end of the monitoring session:

Listing 7.4.21: Order status log exported to CSV.

```
library(IBrokers)

# Connect to Paper Trading TWS
tws <- twsConnect(port = 7497, clientId = 9)

# Define contract for Apple
aapl <- twsEquity("AAPL", exch = "SMART", currency = "USD")

# Create a SELL Limit Order (unlikely to fill immediately)
sellLimitOrder <- twsOrder(
  orderId       = reqIds(tws),
  action        = "SELL",
  totalQuantity = 100,
  orderType     = "LMT",
  lmtPrice      = 250.00   # way above market so it stays open
)

# Place the order
placeOrder(tws, aapl, sellLimitOrder)

# ---- Logging setup ----
orderLog <- data.frame(
  orderId = integer(),
  status = character(),
```

```
  filled = numeric(),
  remaining = numeric(),
  avgFillPrice = numeric(),
  stringsAsFactors = FALSE
)

# Define custom wrapper to log order status
myWrapper <- eWrapper(NULL)

myWrapper$orderStatus <- function(curMsg, msg, ...) {
  row <- data.frame(
    orderId      = as.integer(msg$orderId),
    status       = msg$status,
    filled       = as.numeric(msg$filled),
    remaining    = as.numeric(msg$remaining),
    avgFillPrice = as.numeric(msg$avgFillPrice),
    stringsAsFactors = FALSE
  )
  assign("orderLog", rbind(get("orderLog", envir = .GlobalEnv), row),
         envir = .GlobalEnv)
}

# Listen for updates for ~30 seconds
twsCALLBACK(tws, eWrapper = myWrapper, TIMEOUT = 30)

# Save the log to CSV
csv_file <- "order_status_log.csv"
write.csv(orderLog, file = csv_file, row.names = FALSE)

cat("Order status log exported to:", csv_file, "\n")

# Disconnect
twsDisconnect(tws)
```

Maintain a Single CSV Log Across Sessions

Append order status updates across sessions so that *order_status_log.csv* grows into a continuous trading log rather than being overwritten with each run.

Listing 7.4.22: Maintain a single CSV log across sessions.

```
library(IBrokers)

# Connect to Paper Trading TWS
tws <- twsConnect(port = 7497, clientId = 10)

# Define contract for Apple
aapl <- twsEquity("AAPL", exch = "SMART", currency = "USD")

```

```
# Create a SELL Limit Order (unlikely to fill immediately)
sellLimitOrder <- twsOrder(
  orderId       = reqIds(tws),
  action        = "SELL",
  totalQuantity = 100,
  orderType     = "LMT",
  lmtPrice      = 250.00   # way above market so it stays open
)

# Place the order
placeOrder(tws, aapl, sellLimitOrder)

# ---- Logging setup ----
csv_file <- "order_status_log.csv"

# Initialize orderLog
orderLog <- data.frame(
  orderId      = integer(),
  status       = character(),
  filled       = numeric(),
  remaining    = numeric(),
  avgFillPrice = numeric(),
  stringsAsFactors = FALSE
)

# Custom wrapper to append to log
myWrapper <- eWrapper(NULL)

myWrapper$orderStatus <- function(curMsg, msg, ...) {
  row <- data.frame(
    orderId      = as.integer(msg$orderId),
    status       = msg$status,
    filled       = as.numeric(msg$filled),
    remaining    = as.numeric(msg$remaining),
    avgFillPrice = as.numeric(msg$avgFillPrice),
    stringsAsFactors = FALSE
  )
  assign("orderLog", rbind(get("orderLog", envir = .GlobalEnv), row),
         envir = .GlobalEnv)
}

# Listen for updates for ~30 seconds
twsCALLBACK(tws, eWrapper = myWrapper, TIMEOUT = 30)

# ---- Export & Append ----
if (file.exists(csv_file)) {
  # Append new rows to existing CSV
  existing <- read.csv(csv_file, stringsAsFactors = FALSE)
  combined <- rbind(existing, orderLog)
```

```
  write.csv(combined, file = csv_file, row.names = FALSE)
} else {
  # Create file if it doesn't exist
  write.csv(orderLog, file = csv_file, row.names = FALSE)
}

cat("Order status log appended to:", csv_file, "\n")

# Disconnect
twsDisconnect(tws)
```

Before saving, the script checks whether *order_status_log.csv* already exists. If it does, the existing file is loaded, the new session's `orderLog` is appended, and the combined data is written back. If no file is found, a new one is created. In this way, each session contributes to the same CSV, building a cumulative audit trail of all orders.

Include Timestamps

Include timestamps for each status update in the log, so you can track exactly when TWS sent the message. Every order status update is recorded with a timestamp.

Listing 7.4.23: TSLA SMA(20) crossover backtest.

```
library(IBrokers)

# Connect to Paper Trading TWS
tws <- twsConnect(port = 7497, clientId = 11)

# Define contract for Apple
aapl <- twsEquity("AAPL", exch = "SMART", currency = "USD")

# Create a SELL Limit Order (unlikely to fill immediately)
sellLimitOrder <- twsOrder(
  orderId       = reqIds(tws),
  action        = "SELL",
  totalQuantity = 100,
  orderType     = "LMT",
  lmtPrice      = 250.00    # way above market so it stays open
)

# Place the order
placeOrder(tws, aapl, sellLimitOrder)

# ---- Logging setup ----
csv_file <- "order_status_log.csv"

# Initialize orderLog with a timestamp column
orderLog <- data.frame(
```

```
  timestamp    = character(),
  orderId      = integer(),
  status       = character(),
  filled       = numeric(),
  remaining    = numeric(),
  avgFillPrice = numeric(),
  stringsAsFactors = FALSE
)

# Custom wrapper to log order status with timestamps
myWrapper <- eWrapper(NULL)

myWrapper$orderStatus <- function(curMsg, msg, ...) {
  row <- data.frame(
    timestamp    = format(Sys.time(), "%Y-%m-%d %H:%M:%S"),
    orderId      = as.integer(msg$orderId),
    status       = msg$status,
    filled       = as.numeric(msg$filled),
    remaining    = as.numeric(msg$remaining),
    avgFillPrice = as.numeric(msg$avgFillPrice),
    stringsAsFactors = FALSE
  )
  assign("orderLog", rbind(get("orderLog", envir = .GlobalEnv), row),
         envir = .GlobalEnv)
}

# Listen for updates for ~30 seconds
twsCALLBACK(tws, eWrapper = myWrapper, TIMEOUT = 30)

# ---- Export & Append ----
if (file.exists(csv_file)) {
  existing <- read.csv(csv_file, stringsAsFactors = FALSE)
  combined <- rbind(existing, orderLog)
  write.csv(combined, file = csv_file, row.names = FALSE)
} else {
  write.csv(orderLog, file = csv_file, row.names = FALSE)
}

cat("Order status log (with timestamps) saved to:", csv_file, "\n")

# Disconnect
twsDisconnect(tws)
```

Listing 7.4.23 adds a timestamp column using `Sys.time()`, formatted as `YYYY-MM-DD HH:MM:SS`. Each update from TWS is now logged with the exact local system time at which it was received. As a result, the CSV evolves into a chronological record that spans multiple sessions.

8

Algorithmic Trading: Automation and Monitoring

Champions aren't made in the gyms. Champions are made from something they have deep inside them—a desire, a dream, and a vision.

(Muhammad Ali)

8.1 The Landscape of Algorithmic Trading

Modern financial markets move at a speed and scale that no human trader can consistently match. Screens filled with flashing prices, news headlines, and shifting order books once overwhelmed even seasoned investors. Today, algorithmic trading—sometimes called *algo trading*—has emerged as the solution to this problem. By automating decisions with computer code, traders can act on opportunities faster, more consistently, and with greater discipline than ever before.

At its core, algorithmic trading is the application of pre-defined rules—rooted in data, statistics, and market structure—to the act of trading. Algorithms scan live data feeds, recognize signals, and execute trades in milliseconds. They embody three building blocks:

- **Signals** – the conditions that trigger action, such as a moving-average crossover or a statistical anomaly.
- **Execution logic** – how the order is placed (market, limit, bracket, or stop).
- **Risk management** – capital allocation, stop losses, and drawdown controls.

Yet the promise of algorithmic trading is often misunderstood. It is not a get-rich-quick scheme. Success requires mastery of coding, data analysis, backtesting, and market microstructure. Losing streaks and strategy decay are inevitable; surviving them requires robust risk management and psychological discipline.

8.1.1 From Strategies to Systems

Strategies can be as simple as a moving average crossover or as complex as machine-learning models parsing streaming news. For instance, one design—the *EMA-assisted reversal system*—uses a blend of momentum indicators, RSI confirmations, and a 100-period exponential moving average to capture turning points in volatile, range-bound markets. Other strategies exploit arbitrage gaps across exchanges, ETF rotation signals, or volatility spikes in options.

But profitable strategies alone are not enough. A trading system is more than a single set of rules; it is an engineered framework built for *durability, modularity, and control.* A well-designed system separates analyzers (models that interpret market data) from strategies (decision-making logic), employs robust risk guards, and can recover automatically from failures. Modern frameworks emphasize modular architecture—where input data, analyzers, and strategies operate independently yet interconnect seamlessly—allowing traders to scale across markets, timeframes, and brokers.

8.1.2 Building for Resilience

One of the hardest truths is that most traders abandon good systems too early. Even statistically sound strategies face drawdowns, flat periods, or painful losing streaks. The real challenge lies not in the math, but in *holding the system* when discomfort sets in. Traders who succeed optimize not just for returns but for psychological stability—favoring risk-adjusted measures like the MAR ratio (return divided by maximum drawdown) to ensure they can stay invested.

Resilience also comes from diversification. A single algorithm will eventually break under certain market conditions. A portfolio of uncorrelated strategies—trend following, mean reversion, short-side plays, and volatility systems—creates a buffer. When one approach falters, another cushions the impact, making the overall process sustainable.

8.1.3 Tools, Education, and the Roadmap Ahead

For the aspiring algo trader, the roadmap blends *technical skills* (Python, R, C++, and APIs), *quantitative foundations* (statistics, econometrics, optimization), and *behavioral awareness* (psychology and discipline). Foundational texts—from Ernest Chan's practical handbooks to John Hull's *Options, Futures, and Other Derivatives*—remain essential for grounding strategies in theory and application.

The future of algorithmic trading, particularly for retail traders, is moving toward *open-source, self-hosted frameworks* that provide transparency and regulatory compliance while empowering individuals to build and own their trading stack. Regulators such as SEBI now explicitly recognize the "tech-savvy algo trader"—someone who codes, hosts, and manages their own systems within defined guardrails. Platforms like Interactive Brokers TWS and the TWS API exemplify this shift, offering a compliant bridge for those who prefer to automate.

8.1.4 The Reality of Success in Algorithmic Trading

Algorithmic trading is not magic, nor is it easy. It is a craft that demands sustained learning, rigorous testing, and constant adaptation. Markets evolve, and so must systems. But with the right mindset—structured design, disciplined execution, and realistic expectations—algorithmic trading can be more than just a competitive edge. It can be a path to systematic, sustainable participation in global markets. The goal is not only to understand how algorithms trade, but also to recognize the deeper truth: in algorithmic trading, success belongs not to the fastest coder or the most complex model, but to the trader who builds systems resilient enough to last.

8.2 Designing and Implementing a Trading Strategy

Designing and implementing a trading strategy requires more than intuition; it demands structure, discipline, and the systematic application of quantitative methods. Algorithmic trading has reshaped this process by enabling traders to transform abstract ideas into executable models that operate with speed, precision, and consistency. A strategy begins with a concept—whether moving average crossover, trend following, mean reversion, or statistical arbitrage—and is then refined through data analysis, backtesting, and risk management protocols.

For retail traders, the design phase is an exercise in clarity: defining objectives, specifying rules, and translating them into code. Implementation, in turn, is the bridge from theory to practice, involving platform selection, broker integration, and live execution. Unlike discretionary trading, where emotion often intrudes, algorithmic systems execute pre-defined logic, ensuring discipline and repeatability.

Listing 8.2.1 is an R trading script setup using the IBrokers package to connect to Interactive Brokers' Trader Workstation (TWS).

Listing 8.2.1: AAPL 10-20 simple moving average (SMA) crossover strategy.

```
# ---- Setup ----
library(IBrokers)

# Connect to TWS / IB Gateway
tws <- twsConnect(port = 7497, clientId = 1)

# Use concrete exchange (avoid SMART routing errors 200)
security <- twsEquity("AAPL", exch = "NASDAQ", currency = "USD")

# Parameters
sma_len1 <- 10   # short SMA length
sma_len2 <- 20   # long SMA length
toydata <- data.frame()
currentPosition <- 0
max_pos <- 5

# ---- Logging paths ----
log_dir <- file.path(getwd(), "logs")
dir.create(log_dir, showWarnings = FALSE)

orders_file <- file.path(log_dir, "orders.csv")
pnl_file    <- file.path(log_dir, "pnl.csv")

# Ensure clean disconnect at exit
on.exit({
  if (!is.null(tws)) {
    twsDisconnect(tws)
    message("Disconnected from TWS")
  }
}, add = TRUE)
```

```
2 -1 2104 Market data farm connection is OK:hfarm
2 -1 2104 Market data farm connection is OK:eufarmnj
2 -1 2104 Market data farm connection is OK:cashfarm
2 -1 2104 Market data farm connection is OK:usfuture.nj
2 -1 2104 Market data farm connection is OK:usfuture
2 -1 2104 Market data farm connection is OK:jfarm
2 -1 2104 Market data farm connection is OK:usfarm.nj
2 -1 2104 Market data farm connection is OK:usfarm
2 -1 2104 Market data farm connection is OK:usopt
2 -1 2106 HMDS data farm connection is OK:euhmds
2 -1 2106 HMDS data farm connection is OK:cashhmds
2 -1 2106 HMDS data farm connection is OK:fundfarm
2 -1 2106 HMDS data farm connection is OK:ushmds
2 -1 2158 Sec-def data farm connection is OK:secdefil
```

This output is a status report from Interactive Broker's, showing the health of its market data farm connections.

Listing 8.2.1, Line 2 loads **IBrokers**, which provides R functions for connecting to and trading through Interactive Brokers' API. Line 5 opens a connection to TWS on `port 7497` (default for paper trading). `clientId = 1` identifies this R session. Using unique IDs avoids conflicts if multiple scripts or apps connect simultaneously. Line 8 creates a contract object for Apple stock (AAPL), that uses the NASDAQ exchange instead of SMART routing. This avoids "Error 200: No security definition has been found" messages that sometimes occur when SMART routing is ambiguous. Lines 11–15 define the strategy parameters. Lines 18–22 creates a sub-directory called `logs/` inside the working directory. The `orders_file` is a CSV file for recording all orders placed. The `pnl_file` is a CSV file for recording profit & loss over time. This ensures persistence across runs and provides an audit trail. Lines 25–30 ensures that when the script stops (normally or due to an error), it safely disconnects from TWS. They also prevent "zombie connections" that can accumulate if sessions are not closed properly.

Listing 8.2.2: AAPL 10-20 SMA crossover strategy.

```
# ---- Trading loop ----
while(TRUE) {

  # Request market data snapshot
  tick <- reqMktData(
    tws,
    Contract = security,
    tickGenerics = "236" # Last price tick
  )

  # Extract last price safely
  last_price <- suppressWarnings(as.numeric(tick$last))
  if (is.na(last_price)) {
    message("No last price received.")
    Sys.sleep(1)
    next
  }

  # Append to data log
  toydata <- rbind(toydata, data.frame(Time = Sys.time(), Last = last_
    price))
  toydata <- unique(toydata)

  # Wait until enough data
  if (nrow(toydata) < sma_len2) {
    message("Not enough data yet: ", nrow(toydata), " rows")
    Sys.sleep(1)
    next
```

```
}

# Compute SMAs
short_sma <- mean(tail(toydata$Last, sma_len1), na.rm = TRUE)
long_sma  <- mean(tail(toydata$Last, sma_len2), na.rm = TRUE)

# Generate trading signal
action <- ""
if (short_sma > (long_sma + 0.005)) action <- "BUY"
if (short_sma < (long_sma - 0.005)) action <- "SELL"

quantity <- 1

if (action != "") {
  message("Signal: ", action, " | Position: ", currentPosition)

  # Position risk check
  if (abs(currentPosition +
          (action == "BUY")  * quantity -
          (action == "SELL") * quantity) < max_pos) {

    orderId <- as.numeric(reqIds(tws))

    toyorder <- twsOrder(
      orderId       = orderId,
      action        = action,
      orderType     = "MKT",
      totalQuantity = quantity,
      transmit      = TRUE
    )

    # Place order
    placeOrder(tws, security, toyorder)

    # Update position
    currentPosition <- currentPosition +
      (action == "BUY")  * quantity -
      (action == "SELL") * quantity

    # ---- Log order to CSV ----
    order_log <- data.frame(
      Time      = Sys.time(),
      OrderId   = orderId,
      Action    = action,
      Quantity  = quantity,
      LastPrice = last_price,
      stringsAsFactors = FALSE
    )
    write.table(
```

```
        order_log,
        file = orders_file,
        sep = ",",
        append = TRUE,
        col.names = !file.exists(orders_file),
        row.names = FALSE
      )

      # ---- Log PnL placeholder (expand later) ----
      pnl_log <- data.frame(
        Time      = Sys.time(),
        Position  = currentPosition,
        MarketVal = currentPosition * last_price,
        stringsAsFactors = FALSE
      )
      write.table(
        pnl_log,
        file = pnl_file,
        sep = ",",
        append = TRUE,
        col.names = !file.exists(pnl_file),
        row.names = FALSE
      )
    }

  } else {
    message(
      "No action | SMA(short)=", round(short_sma, 3),
      " SMA(long)=", round(long_sma, 3),
      " | Position=", currentPosition
    )
  }

  Sys.sleep(1)
}
```

```
<20251001 21:17:00.952120> id=1 symbol=AAPL lastTimestamp: 1759342620
<20251001 21:17:00.954047> id=1 symbol=AAPL lastPrice: 256.63
<20251001 21:17:00.955125> id=1 symbol=AAPL lastSize: 1
<20251001 21:17:00.956049> id=1 symbol=AAPL Volume: 264904
<20251001 21:17:00.956955> id=1 symbol=AAPL highPrice: 258.79
<20251001 21:17:00.957870> id=1 symbol=AAPL lowPrice: 254.93
<20251001 21:17:00.958775> id=1 symbol=AAPL closePrice: 254.63
<20251001 21:17:00.959675> id=1 symbol=AAPL openPrice: 254.99
<20251001 21:17:01.312875> id=1 symbol=AAPL bidPrice: 256.63
          bidSize: 1
```

```
<20251001 21:17:01.314445> id=1 symbol=AAPL askPrice: 256.65
          askSize: 4
<20251001 21:17:01.315923> id=1 symbol=AAPL bidSize: 1
<20251001 21:17:01.347706> id=1 symbol=AAPL askSize: 4
<20251001 21:17:01.530148> id=1 symbol=AAPL shortable: 3.0
<20251001 21:17:01.531703> id=1 symbol=AAPL <default size> 6 1 89
          87590749
<20251001 21:17:01.533407> id=1 symbol=AAPL lastTimestamp: 1759342621
...
```

The output above displays the latest available market data for AAPL (Apple Inc.). `timestamp on October 1, 2025, 21:17:01` provides a snapshot of Apple stock's trading activity in terms of price levels, volume, bid/ask spread, and short availability at that moment.

Listing 8.2.2, Line 2 initiates an infinite trading loop. The script runs indefinitely (`while(TRUE)`). Each cycle executes once per second (`Sys.sleep(1)`). Lines 5–9 `Asks` IBKR for a market data snapshot of the defined security (e.g., AAPL stock). `tickGenerics = "236"` requests the last traded price. Lines 12–17 extract the last price. It converts the reported price to numeric. If no price is available (`NA`), the loop waits and retries. Lines 20–21 `appends` a new row with `timestamp` and `last price` to the running dataset `toydata`, and removes duplicates to avoid multiple identical rows. Lines 24–28 ensures there's enough historical data before calculating SMAs. It requires at least `sma_len2` rows (the longer SMA length).

Lines 31–32 compute SMAs which, are the backbone of the SMA crossover strategy. Lines 35–42 generate the trading signals. If the short SMA rises above long SMA (plus a small threshold), `signal = BUY`. If the short SMA falls below long SMA (minus threshold), `signal = SELL`. However, if neither condition holds, no action is taken (HOLD). Lines 45–57 provide risk management by setting position limits. This ensures the position after the new order won't exceed `max_pos`. `currentPosition` tracks open exposure (long/short); which, prevents runaway accumulation of shares/contracts. Line 60 handles order creation and submission. Lines 63–65 update position state by keeping an internal accounting of current holdings.

Order logging is implemented in Lines 68–83. Each order is logged into `orders.csv` which, records `timestamp`, `orderID`, `action`, `size`, and `price`. This ensures persistence across sessions.

PnL logging is implemented in Lines 86–100. `pnl_log` and `write.table` logs a simple snapshot of profit and loss into `pnl.csv`. Currently, it only computes `market value = position` × `price`. However, this could be expanded to include realized/unrealized PnL.

No-action logging is implemented in Lines 102–111. If no trade occurs, it prints the SMA values and current position to the console. This helps with debugging and monitoring.

9

QuantRoom Videos and Scripts

> A dream is not something that you wake up from, but something that wakes you up.
>
> *(Charlie Hedges)*

9.1 Introduction

The *QuantRoom* YouTube channel (`https://www.youtube.com/@quantroom`) serves as a practical hub for traders and developers interested in algorithmic trading, quantitative analysis, and brokerage API integration. The channel blends tutorial-style videos, live coding sessions, and walkthroughs of trading strategies, focusing especially on Interactive Brokers (IBKR), Schwab API workflows, R/Python scripting, and real-time trading infrastructure. Its content bridges theory and practice, guiding viewers from basic setup to sophisticated trading system design. Alongside videos, *QuantRoom* often shares scripts, snippets, and coding best practices, making it a valuable companion for readers who want both conceptual clarity and hands-on implementation.

9.2 Interactive Brokers Videos and Scripts

Below is a curated selection of videos from the *QuantRoom* channel featuring references to IBKR. Each entry provides the video title along with its URL. You can also locate these videos by searching the titles directly in the *QuantRoom* search bar.

1. **How To Set Up Live Trading Environment With Interactive Brokers**
   `https://www.youtube.com/watch?v=H4jWZnXHi98`

2. **How to Use Interactive Brokers TWS 2025 | Chart, Order Entry**
   `https://www.youtube.com/watch?v=fFSKgXZQ7Wg`

3. **How To Use Interactive Brokers Effectively (2025)**
   `https://www.youtube.com/watch?v=gLLIJ8NeLuI`

4. **Introduction to Charts – IBKR Desktop**
   `https://www.youtube.com/watch?v=cBNcjeL6upk`

5. **IBKR's Client Portal – Options Trading**
   `https://www.youtube.com/watch?v=r3X1bfq4lzA`

6. **Option Algorithmic Trading with Interactive Brokers**
   `https://watch?v=zNQY5Krv8aA&list=PLvppnAE-9cZWsWyc30xtzKNIzSmGCJB1w`

7. **Algorithmic Trading Support/Resistance Levels on Interactive Brokers using Bracket Orders**
   `https://www.youtube.com/watch?v=i18EqtQERuU&t=32s`

8. **Covered Call Options Trading with Interactive Brokers**
   `https://www.youtube.com/watch?v=99DDK9DDHHQ&t=2s`

9. **Interactive Brokers API - How to Get Historical Data for Futures, Options, and Stocks**
   `https://www.youtube.com/watch?v=UHGNua_Dxlw&t=3s`

10. **Interactive Brokers API - Stock Fundamental Data**
    `https://www.youtube.com/watch?v=EQd0YVI7otc&t=2s`

11. **Interactive Brokers API - How To Process Messages For Real-Time Market Data**
    `https://www.youtube.com/watch?v=1oE_mP9NLAk&t=2s`

12. **Algorithmic Trading Performance for the First Week**
    `https://www.youtube.com/watch?v=f3H0ydG2aiE&t=43s`

13. **Interactive Brokers API - Streaming Real-Time Bars for Futures, Options, & Stocks**
    `https://www.youtube.com/watch?v=RHpBo-xNCtE&t=25s`

14. **Interactive Brokers API - Market Scanners**
    `https://www.youtube.com/watch?v=cDQxm9qBDvw&t=4s`

9.3 Charles Schwab (Trader) API Videos and Scripts

Below is a curated selection of video titles from the *QuantRoom* channel featuring references to Charles Schwab (Trader) API. Each entry provides the video title. You can locate the URL links to these videos by searching the titles directly in the *QuantRoom* search bar.

1. **Getting Started**
2. **Authentication**
3. **Getting Option Chains (Single and Spreads)**
4. **Getting Historical Data for Stocks, Futures, Indices and Forex**
5. **Miscellaneous Market Data Wrappers**
6. **Accessing and Retrieving Account Information**
7. **How to Place, Replace, and Cancel Single Equity/Option Orders**
8. **Complex Orders — Option Spreads, Trailing Stops, and Trigger Orders**
9. **Schwab API Function Library**
10. **How to Create an Intraday Stock Database Using the Schwab Trader API**
11. **Algorithmic Trading Using the Charles Schwab API — Trading Intraday Stock Ranks**
12. **Getting All Stock Fundamentals Via the Trader API**
13. **Algorithmic Trading Technical/Fundamental Strategy Using the Trader API**
14. **Making R-Code Executables (Mac/Win) — Simplifying Trader API**

From a strategic standpoint, maintaining dual API integration across IBKR and Schwab Trader API provides algorithmic traders with substantial flexibility and resilience. The socket-based architecture of *IBKR* excels in low-latency environments where event-driven execution and rapid order routing are critical. Conversely, the RESTful design of the *Schwab Trader API* offers stateless, cloud-compatible access ideally suited for scheduled tasks, batch analytics, and algorithmic portfolio management. By leveraging both interfaces within a unified R or Rcpp workflow, traders can distribute computational and execution workloads dynamically—using IBKR for real-time signal deployment while utilizing Schwab for broader market data aggregation, historical retrieval, and intraday strategy orchestration. This multi-broker architecture mitigates operational risk, enhances data redundancy, and strengthens the robustness of automated trading infrastructure across heterogeneous environments.

Why Videos and Scripts Matter

The dual offering of videos and scripts creates a learning environment that appeals to both visual and hands-on learners. Videos provide step-by-step demonstrations, making complex topics such as order routing, API setup, or real-time data streaming easier to follow. Scripts, on the other hand, allow viewers to move beyond passive watching by experimenting directly with the scripts in their own environments. This combination accelerates skill acquisition: concepts introduced in tutorials are immediately reinforced through implementation, while scripts serve as reusable building blocks for more advanced systems. For traders and developers, this integration of visual explanation with executable code significantly lowers the barrier from theory to practice, ensuring not only understanding but also practical application.

Conclusion

The *QuantRoom* channel contains a *dedicated set of tutorials* on Interactive Brokers, ranging from platform usage (TWS, Desktop, Client Portal) to API integration using *IBrokers*. These resources are designed to help traders integrate IBKR into both manual and algorithmic trading workflows. In parallel, an extensive series of sessions is devoted to the *Charles Schwab Trader API*, covering authentication, data access, order management, and automation pipelines. This dual focus on *IBKR* and *Schwab Trader API* illustrates a unified framework for API-driven execution, enabling algorithmic traders to seamlessly alternate between brokers, diversify infrastructure risk, and extend strategy deployment capabilities. This integration of explanation and execution not only makes both IBKR and Schwab more accessible but also empowers traders to design, test, and refine their own manual and algorithmic strategies with confidence (Table 9.1).

TABLE 9.1: Comparison of core features and architectural design between the Interactive Brokers (IBKR) API and the Charles Schwab Trader API. Both interfaces enable programmatic access to market data and order execution; however, Schwab's REST-based design favors stateless algorithmic pipelines, whereas IBKR's socket-based API allows for persistent, low-latency trading systems.

Feature	IBKR API	Charles Schwab API
Primary Interface	`IBrokers` (R), `ib_insync` (Python), `C++ / Java TWS API`	REST/JSON endpoints with OAuth2 authentication; R and Python wrappers under development
Connection Mode	Socket-based, persistent client connection via TWS or Gateway	HTTPS-based REST service; tokenized session with short-lived access tokens
Authentication Flow	Local host/port + clientID authorization	OAuth2: redirect URI, client ID/secret, refresh token
Market Data Access	Real-time tick, level I/II, historical bars, contracts	Delayed/real-time quotes, option chains, historical OHLC via REST
Order Management	Full suite (market, limit, stop, combo, bracket, OCA)	Equity and option orders (market, limit, stop, trailing, spread)
Streaming / WebSocket	Supported via `ib_insync` event loop or EWrapper callbacks	Limited streaming; beta WebSocket for quote updates
Historical Data Granularity	1 sec to 1 day bars; flexible look-back windows	Minute, hourly, and daily bars; capped request frequency
Algorithmic Strategy Integration	Seamless integration with Rcpp, Python, and Java; supports low-latency event-driven trading	Script-driven REST queries for batch or periodic execution; suited for swing and intraday analytics
Rate Limits	Moderate (max 50 req/s typical)	Strict per-minute quota (approx. 120 calls/min per app)
Paper / Demo Environment	Full demo accounts (IB Gateway, TWS)	Separate sandbox endpoint provided via developer portal
Cross-Platform Compatibility	Windows, macOS, Linux	Cloud-agnostic; REST via HTTPS (Mac/Win/Linux compatible)
Integration Languages	R, Python, C++, Java, Excel VBA	R, Python, Node.js, cURL, Bash

Appendix A

Comparison of C++ TWS API and R IBrokers Package

Introduction

This appendix provides a comparative framework for integrating the Interactive Brokers (IB) Trader Workstation (TWS) C++ API with R through the **IBrokers** package and **Rcpp**. It begins by showing that while **IBrokers** offers a convenient high-level R interface for prototyping and statistical modeling, it has limitations in event-driven trading, advanced order handling, and low-latency data processing. To address these, **Rcpp** can expose the full functionality of the native C++ API directly within R. The appendix includes: (i) a functional mapping table that aligns TWS C++ API calls with their **IBrokers** counterparts, covering connection management, market data, historical queries, real-time bars, orders, accounts, executions, contracts, and error handling; and (ii) a comparative feature analysis highlighting differences in ease of use, performance, API coverage, event handling, and customizability. Together, these materials position **IBrokers** as best suited for research and prototyping, while **Rcpp**-enabled C++ integration extends R into a platform capable of supporting robust, event-driven, and latency-sensitive trading systems.

Motivation

Algorithmic trading demands both flexibility and efficiency. Researchers and developers working in R often prefer the ease of prototyping and statistical modeling that **IBrokers** provides. However, advanced trading systems—particularly those that rely on event-driven callbacks, custom order types, or low-latency data handling—require direct access to the native C++ TWS API. By using **Rcpp**, we can integrate these capabilities into the R environment without sacrificing usability.

Limitations of `IBrokers`

Although **IBrokers** is stable and well documented, several limitations restrict its applicability in high-performance scenarios:

- Limited exposure to advanced IBKR order types and event handlers.
- Dependency on synchronous request/response workflows, which are less suitable for real-time trading.
- Reduced flexibility for handling large data streams and custom trading logic at the low level.

These constraints highlight the need for a complementary approach that leverages the underlying C++ API directly.

Advantages of Integrating C++ via Rcpp

Rcpp provides a natural bridge between R and C++, making it possible to wrap native API calls and expose them as R functions. This integration offers several advantages:

- **Performance:** Critical routines can run in C++ for speed while maintaining R as the orchestration layer.
- **Extensibility:** Direct access to the TWS C++ API allows the use of advanced features not covered by `IBrokers`.
- **Seamless workflow:** Analysts can continue to use R for statistical modeling, visualization, and portfolio analysis while delegating execution to the C++ layer.

Core Differences

- **C++ Core:** Implements the event-driven architecture (via `EClient`, `EWrapper`, `EReader`). Direct access to sockets, threading, and message parsing. Suitable for high-performance, production-level integration.
- **R Wrappers:** Provide functional entry points (e.g., `reqMktData()`, `placeOrder()`) that internally manage connections, message decoding, and error handling. Focus is on ease of use, not raw performance.
- **Design Goal:** C++ offers maximum flexibility and speed; R wrappers trade some of this control for simplicity and accessibility within statistical workflows.

Functional Mapping

Functionality	C++ (TWS API)	R (IBrokers)
Connection Management	EClientSocket, EReader, EReaderSignal	twsConnect(), twsDisconnect(), isConnected()
Market Data (Level I)	EClient::reqMktData() returns tick-by-tick data via tickPrice() callbacks	reqMktData() streams quotes into R data frames via eWrapper.data()
Market Depth (Level II)	EClient::reqMktDepth() with updateMktDepth() callback	reqMktDepth() with optional CSV logging wrappers (eWrapper.MktDepth.CSV)
Historical Data	EClient::reqHistoricalData() returns bars via historicalData()	reqHistoricalData() returns OHLCV data as xts objects
Real-Time Bars	EClient::reqRealTimeBars() delivers 5-second bars via callbacks	reqRealTimeBars() returns 5-second OHLCV in R time-series
Order Management	EClient::placeOrder(), cancelOrder()	placeOrder(), cancelOrder() with R objects (twsOrder, twsContract)
Account Data	EClient::reqAccountUpdates() + callback handlers	reqAccountUpdates() populates R lists or data frames
Executions	EClient::reqExecutions() with execDetails() callback	reqExecutions() returning R objects (twsExecution)
Contract Details	EClient::reqContractDetails() returns contract descriptions	reqContractDetails() wraps contracts in twsContract() R objects
Scanner	EClient::reqScannerSubscription() with callbacks	Scanner(), ScannerSubscription() in R
Error Handling	EWrapper::error() callback	errorHandler.R, exceptions raised into R console

Strengths of Each Layer

C++ Core

- High performance and fine-grained control over sockets and threading.
- Lowest latency, suitable for HFT and execution-sensitive strategies.
- Complete access to every IB API call and experimental endpoints.

R Wrappers

- Rapid prototyping and ease of use within RStudio workflows.
- Tight integration with R's financial ecosystem (`quantmod`, `xts`, `PerformanceAnalytics`).
- Simplified event handling through `eWrapper` objects and data frames.

Table A.1 clearly signals when to use IBrokers alone (e.g., prototyping, education) and when to combine it with Rcpp + C++ (e.g., production trading, advanced API access).

Intended Audience

This guide is intended for:

- Quantitative researchers and developers building custom trading systems in R.
- Academics who wish to extend their econometric and statistical models into live trading without leaving the R ecosystem.
- Practitioners seeking to combine the rapid prototyping of **IBrokers** with the full power of the TWS C++ API.

Contribution of this Guide

By presenting a comparison between **IBrokers** and the direct use of the TWS C++ API through **Rcpp**, this guide contributes:

1. A clear articulation of when to use **IBrokers** alone versus when to extend with **Rcpp**.
2. Example workflows that combine both approaches for maximum efficiency.
3. A roadmap for integrating low-level API calls into R-based research and production environments.

Ultimately, the integration of C++ with R through **Rcpp** should be viewed not as a replacement of **IBrokers**, but as an augmentation that expands the possibilities for serious algorithmic trading in the R ecosystem.

TABLE A.1: Comparison of `IBrokers` and C++ API integration via `Rcpp`.

Feature	IBrokers (R)	C++ API via Rcpp
Ease of Use	High-level R interface; suitable for rapid prototyping and basic trading tasks	Requires C++ knowledge and Rcpp integration; steeper learning curve but more flexible
Performance	Adequate for most research and low-frequency trading; synchronous request/response model	Optimized for speed; can handle low-latency, high-throughput event-driven workflows
API Coverage	Provides access to common market data requests, simple orders, and account queries	Full coverage of Interactive Brokers TWS C++ API, including advanced order types and events
Event Handling	Limited; relies on polling and callbacks within R's event loop	Native event-driven architecture with fine-grained control over order states, market data, and errors
Customizability	Restricted to the functions implemented in IBrokers	Highly extensible: developers can expose any C++ method directly into R via Rcpp
Integration	Seamless with R modeling, visualization, and packages	Requires custom wrappers, but once integrated, combines C++ execution with R analytics
Use Case	Best for research, prototyping, educational use, and straightforward trading strategies	Best for production systems, advanced order logic, and latency-sensitive applications

Appendix B

R C++ Application Programming Interface

B1. Overview

R offers an application programming interface (API) that allows developers to extend its functionality and interface with other languages or libraries. The R application programming interface is defined in the header files bundled with each installation and thoroughly documented in the Writing R Extensions manual (R Development Core Team, Version 4.5.0 [2025-04-11]). The R Core Team consistently advises developers to use only the public API, noting that internal, undocumented routines may be modified or removed without prior notice.

Several works provide detailed discussions of the R API and its applications. Chambers (2008) remains the definitive reference in the context of Programming with Data, and he also presents a more recent perspective, Chambers (2016).

R provides two principal interfaces for extending its functionality with compiled code: the legacy `.C()` interface and the modern `.Call()` interface.

The `.C()` interface, which dates back to the earliest versions of R and S, supports only pointers to basic C data types (such as `double*`, `int*`, and `char*`). This makes it straightforward but limited, offering no way to manipulate complex R objects or return structured data efficiently.

By contrast, the `.Call()` interface operates directly on `SEXP` objects (R's internal S-expression type), which represent every R object: `vectors`, `lists`, `functions`, `environments`, and more. This provides full access to R's memory management and object system from compiled code, allowing developers to construct, inspect, and modify arbitrary R objects within C or C++ routines.

Modern R extensions, including all packages built with **Rcpp**, rely exclusively on `.Call()`. Rcpp automatically generates the `.Call()` interface code that connects R and C++ functions, handling the conversion between R objects and native C++ types safely and efficiently. This abstraction layer allows developers to write idiomatic C++ without dealing directly with `SEXP` pointers, while preserving the underlying performance benefits and memory safety mechanisms of R's C API.

Rcpp builds upon R's underlying API to offer a complementary and more accessible interface for extending the language. By leveraging the abstractions and type-safety mechanisms of modern C++, **Rcpp** provides a more coherent and intuitive programming model, one that aligns closely with how R users work with data objects.

In summary (Table B.1), we have used the current terminology found in `R 4.x` and recent **Rcpp** documentation (`1.0.12+`). Furthermore, it more accurately reflects modern **Rcpp** workflows by noting the central role of automatic interface generation, particularly via the `[[Rcpp::export]]` attribute. Finally, we have clarified the status of the `.C()` interface, correctly identifying it as *deprecated* for new development, even though it remains supported for backward compatibility.

TABLE B.1: Comparison of legacy and modern methods for extending R with compiled code via **Rcpp**. Legacy workflows based on the inline package have been superseded by the attribute-based sourceCpp() interface, which automates compilation, linking, and registration.

Topic	Legacy (pre-2015)	Modern (2025)
Interactive compilation	`inline::cfunction()` / `inline::cxxfunction()`	`Rcpp::sourceCpp()`
Integration with Rcpp	`plugin="Rcpp"`	`[[Rcpp::export]]`
Function registration	Manual or `inline`	Automatic via `compileAttributes()`
Recommended usage	Deprecated	Fully supported and preferred

B2. A First Compilation

Following Eddelbuettel (2013), we consider the example shown in Listing B2.1, and assume that both the fibonacci function and the wrapper have been saved in a file named `fibCpp.cpp` (Listing B1.1). Then, on a 64-bit macOS computer with **Rcpp** installed in a standard location, we can compile it via the example shown in Listing B2.2.

Listing B2.1. **Rcpp** C++ code file `fibCpp.cpp`.

```
#include <Rcpp.h>
using namespace Rcpp;

// [[Rcpp::export]]
int fibCpp(int n) {
  if (n == 0) return 0;
  if (n == 1) return 1;
  int a = 0, b = 1, temp;
```

```
  for (int i = 2; i <= n; i++) {
    temp = a + b;
    a = b;
    b = temp;
  }
  return b;
}
```

Listing B2.2. Commands for automatically building `fibCpp.cpp` using `sourceCpp("fibCpp.cpp")`.

```
library(Rcpp)
sourceCpp("fibCpp.cpp")
```

Listing B2.2 invokes R's C++ toolchain (clang++ on macOS, g++ on Linux/Windows), compiles `fibCpp.cpp` into a shared object (i.e., `fibCpp.so` or `fibCpp.dll`), loads it automatically into the R session, and registers `fibCpp()` as an R function (Figure B2.1).

Figure B2.1. `fibCpp` function registered in R's Global Environment.

The Inline Package

The **inline** package and its functions `cfunction()` and `cxxfunction()` were popular in early **Rcpp** development (circa 2010–2015) for interactive compilation. These functions still exist, but they're now considered legacy tools. The **inline** package is no longer actively maintained, and it's not recommended for new projects or reproducible package development. CRAN and **Rcpp** documentation now emphasize `Rcpp::sourceCpp()` as the standard supported alternative.

Extending R with compiled code requires an effective method for compiling, linking, and dynamically loading code. While embedding routines within an R package is ideal for the long term, it can be cumbersome for exploratory work.

Modern **Rcpp** development uses the `Rcpp::sourceCpp()` interface, which allows for straightforward compilation and loading of C++ source files directly from R. This function automatically handles compiler flags, headers, linking, and function registration, eliminating the need for manual `Makefile` setup.

Previous workflows relied on the **inline** package and directives like "**Rcpp**" to manage include paths and linking options. While still functional for legacy code, these methods have been replaced by attribute-based exports (i.e., `[[Rcpp::export]]`) and automatic registration through `Rcpp::compileAttributes()`.

Using `Includes`

The following code example is from the **Rcpp** FAQs. It originated from a user query on the **Rcpp** mailing list (Eddelbuettel (2013)). It defines a simple templated class that computes the square of its argument, supplied within an `include` file. The main function then demonstrates the flexibility of this template by applying it to two different data types.

Listing B2.3 presents the modernized, fully runnable **Rcpp** example rewritten to use `sourceCpp()` (the preferred method since `inline::cxxfunction()` is now deprecated):

Listing B2.3. Save this code as `square_template.cpp`.

```
#include <Rcpp.h>
using namespace Rcpp;

// Template class that squares its argument
template <typename T>
class square {
public:
    T operator()(T t) const {
        return t * t;
    }
};

// [[Rcpp::export]]
DataFrame fun(double xs, int is) {
    square<double> sqdbl;
    square<int> sqint;

    DataFrame df = DataFrame::create(
        Named("x") = sqdbl(xs),
        Named("i") = sqint(is)
    );

    return df;
}
```

Listing B2.4. Save this code and run it in RStudio using:

```
library(Rcpp)
sourceCpp("square_template.cpp")

fun(2.2, 3L)
```

The directive `#include <Rcpp.h>` grants access to **Rcpp**'s C++ interface, enabling seamless integration between C++ and R. Within the code, the templated class `square<T>` defines a generic `operator()` that returns the square of its argument, t^2. The function `fun()` then creates instances of `square<double>` and `square<int>` to handle both double and integer inputs, applies the squaring operation to each value, and compiles the results into an R `DataFrame`. Finally, the `[[Rcpp::export]]` attribute ensures that the function is automatically exposed to R and can be invoked directly after compilation using `sourceCpp()`.

Using Plugins

Plugins provide a flexible mechanism that allows packages using **Rcpp** to specify any additional information required for compilation and linking. This may include extra header files and include directories, as well as library names and their corresponding locations needed during the build process.

The code below (Listing B2.5) demonstrates the use of a plugin through a concise, modern example—a fully runnable `sourceCpp()` implementation of the classic **RcppArmadillo** regression model. This version adopts the current **RcppArmadillo** export mechanism and employs clean, idiomatic C++14 syntax for clarity and efficiency.

Listing B2.5. Save this code and run it in RStudio using:

```
library(Rcpp)
library(RcppArmadillo)
sourceCpp("ols_fit.cpp")

# Example data
set.seed(123)
X <- cbind(1, rnorm(10))          # intercept + one predictor
y <- 1 + 2 * X[,2] + rnorm(10)

ols_fit(y, X)
```

Listing B2.6. Save this code and run it in RStudio using:

```
#include <RcppArmadillo.h>
using namespace Rcpp;

// [[Rcpp::depends(RcppArmadillo)]]

// [[Rcpp::export]]
List ols_fit(const NumericVector& ys, const NumericMatrix&
Xs) {
    // Convert R objects to Armadillo structures
```

```
    int n = Xs.nrow(), k = Xs.ncol();
    arma::mat X(Xs.begin(), n, k, false);
    arma::colvec y(ys.begin(), ys.size(), false);

    // —— Fit OLS: y = X * beta + e ——
    arma::colvec coef = arma::solve(X, y);
    // coefficients
    arma::colvec res  = y - X * coef;
    // residuals

    // —— Residual variance (sigma^2) ——
    double s2 = arma::dot(res, res) / (n - k);

    // —— Standard errors ——
    arma::colvec se = arma::sqrt(s2 * arma::diagvec
       (arma::inv(X.t() * X)));

    // —— Return results ——
    return List::create(
        Named("coef") = coef,
        Named("se")   = se,
        Named("df")   = n - k
    );
}
```

Discussion

- **Includes:** `#include <RcppArmadillo.h>` provides simultaneous access to both the **Rcpp** and **Armadillo** libraries, enabling seamless integration of R and high-performance C++ linear algebra routines.
- **Conversion:** The Armadillo types `arma::mat` and `arma::colvec` are directly mapped to R's underlying memory via `begin()` pointers. This zero-copy interface allows computations to occur without redundant data duplication, preserving efficiency.
- **Computation:**
  - Coefficients are estimated using the ordinary least squares normal equations:

$$\hat{\beta} = (X'X)^{-1}X'y = \texttt{arma::solve(X, y)}$$

  - Residuals are obtained as:

$$e = y - X\hat{\beta}$$

- The residual variance is computed as:

$$s^2 = \frac{e'e}{n-k}$$

where n is the number of observations and k is the number of regressors.

- The standard errors are derived from:

$$\text{se}(\hat{\beta}) = \sqrt{s^2 \,\text{diag}\left((X'X)^{-1}\right)}$$

- **Output:** The function returns an `Rcpp::List` containing:
  - `coef` — the estimated coefficients,
  - `se` — the corresponding standard errors, and
  - `df` — the residual degrees of freedom, $n-k$.

Bibliography

Amatsuki, Kaito. *AI Options Trading With Python.* 2025. Kindle edition.

Amatsuki, Kaito. *AI Quant Trading.* 2025. Kindle edition.

Amatsuki, Kaito. *Algorithmic Trading With Python.* 2025. Kindle edition.

Amatsuki, Kaito. *Predictive AI Trading.* 2025. Kindle edition.

Ang, Clifford S. *Analyzing Financial Data and Implementing Financial Models Using R.* Second Edition. Cham, Switzerland: Springer, 2021.

Baranasooriya, Melick R. *High-Performance Algorithmic Trading Using AI: Strategies and Insights for Developing Cutting-Edge Trading Algorithms.* (English Edition) India: BPB Publications, 2024.

Barreda, Santiago, and Noah Silbert. *Bayesian Multilevel Models for Repeated Measures Data: A Conceptual and Practical Introduction in R.* New York, NY: Routledge, Taylor & Francis Group, 2023.

Barrett, Tyson. 2025. *data.table: Extension of 'data.frame'.* Version 1.17.8. R package. CRAN. `https://cran.r-project.org/web/packages/data.table/index.html`.

Becker, Marc. 2025. *mlr3: Machine Learning in R - Next Generation.* Version 1.1.0. R package. CRAN. `https://cran.r-project.org/web/packages/mlr3/index.html`.

Biecek, Przemyslaw. 2025. *DALEX: moDel Agnostic Language for Exploration and eXplanation.* Version 2.5.2. R package. CRAN. `https://cran.r-project.org/web/packages/DALEX/index.html`.

Boehmke, Bradley C. *Data Wrangling with R.* Cham, Switzerland: Springer, 2016.

Breiman, Leo. 2001. "Random Forests." *Machine Learning* 45 (1): 5–32.

Bryant, James, and Aloke Mukherjee. *The Future of Finance with ChatGPT and Power BI.* Mumbai, India: Packt Publishing, 2023.

Bürkner, Paul-Christian. 2024. *brms: Bayesian Regression Models using 'Stan'*. Version 2.22.0. R package. CRAN. `https://cran.r-project.org/web/packages/brms/index.html`.

Burns, Steve, and Holly Burns. *Five Moving Average Signals That Beat Buy and Hold.* Gulf Breeze, FL: Stolly Media, LLC, 2017.

Burns, Steve, and Holly Burns. *Moving Averages 101.* Third Edition. Gulf Breeze, FL: Stolly Media, LLC, 2025.

Casalicchio, Giuseppe. 2025. *iml: Interpretable Machine Learning.* Version 0.11.4. R package. CRAN. `https://cran.r-project.org/web/packages/iml/index.html`.

Chan, Ernest P. *Algorithmic Trading: Winning Strategies and Their Rationale.* Hoboken, New Jersey: John Wiley & Sons, Inc., 2013.

Chan, Ernest P. *Machine Trading: Deploying Computer Algorithms to Conquer the Markets.* Hoboken, New Jersey: John Wiley & Sons, Inc., 2017.

Chan, Ernest P. *Quantitative Trading: How to Build Your Own Algorithmic Trading System.* Hoboken, New Jersey: John Wiley & Sons, Inc., 2009.

Chang, Winston. 2025. *shiny: Web Application Framework for R.* Version 1.11.1. R package. CRAN. `https://cran.r-project.org/web/packages/shiny/index.html`.

Collard, Jean-François. *Hands-On Data Analysis in R for Finance.* Boca Raton, FL: CRC Press, Taylor & Francis Group, 2023.

Conlan, Chris. *Automated Trading with R: Quantitative Research and Platform Development.* New York, NY: APRESS, 2016.

Congdon, Peter D. *Bayesian Hierarchical Models With Applications Using R.* Second Edition. Boca Raton, FL: CRC Press, Taylor & Francis Group, 2020.

Congdon, Peter D. *Bayesian Statistical Modeling.* Second Edition. Chichester, West Sussex, England: John Wiley & Sons, Ltd., 2006.

Cotton, Richard. *Learning R.* Sebastopol, CA: O'Reilly Media, Inc., 2013.

Davey, Kevin J. *Building Winning Algorithmic Trading Systems.* Hoboken, New Jersey: John Wiley & Sons, Inc., 2014.

Dancho, Matt. 2025. *tidyquant: Tidy Quantitative Financial Analysis.* Version 1.0.11. R package. CRAN. `https://cran.r-project.org/web/packages/tidyquant/index.html`.

Dancho, Matt M. 2023. *alphavantager: Lightweight Interface to the Alpha Vantage API.* Version 0.13. R package. CRAN. `https://cran.r-project.org/web/packages/alphavantager/index.html`.

Dunis, Christian L., Peter W. Middeleton, A. Karathanaspoulus, and Konstantinos Theofilatos. *Artificial Intelligence in Financial Markets.* London, UK: Palgrave Macmillan, 2017.

Eddelbuettel, Dirk. 2023. *Rblpapi: R Interface to 'Bloomberg'.* Version 0.3.16. R package. CRAN. `https://cran.r-project.org/web/packages/Rblpapi/index.html`.

Eddelbuettel, Dirk. 2025. *RQuantLib: R Interface to the 'QuantLib' Library.* Version 0.4.26. R package. CRAN. `https://cran.r-project.org/web/packages/RQuantLib/index.html`.

Eddelbuettel, Dirk. *Seamless R and C++ Integration with Rcpp.* New York, NY: Springer, 2013.

Edwards, Robert D., John Magee, and W. H. C. Bassetti. *Technical Analysis of Stock Trends.* 11th Edition. New York: Routledge, 2019.

Enos, Jeff. 2020. *strand: A Framework for Investment Strategy Simulation.* Version 0.2.0. R package. CRAN. `https://cran.r-project.org/web/packages/strand/index.html`.

Eubank, Randall L., and Ana Kupresanin. *Statistical Computing in C++ and R.* Boca Raton, FL: CRC Press, 2011.

Falbel, Daniel. 2025. *torch: Tensors and Neural Networks with 'GPU' Acceleration.* Version 0.16.0. R package. CRAN. `https://cran.r-project.org/web/packages/torch/index.html`.

Friedman, Jerome H. 2001. "Greedy Function Approximation: A Gradient Boosting Machine." *Annals of Statistics* 29 (5): 1189–1232.

Friedman, Jerome. 2025. *glmnet: Lasso and Elastic-Net Regularized Generalized Linear Models.* Version 4.1-10. R package. CRAN. `https://cran.r-project.org/web/packages/glmnet/index.html`.

Fryda, Tomas. 2024. *h2o: R Interface for the 'H2O' Scalable Machine Learning Platform.* Version 3.44.0.3. R package. CRAN. `https://cran.r-project.org/web/packages/h2o/index.html`.

Galanos, Alexios. 2025. *rmgarch: Multivariate GARCH Models.* Version 1.4-2. R package. CRAN. `https://cran.r-project.org/web/packages/rmgarch/index.html`.

Galanos, Alexios. 2025. *rugarch: Univariate GARCH Models.* Version 1.5-4. R package. CRAN. `https://cran.r-project.org/web/packages/rugarch/index.html`.

Gerlanc, Daniel. 2023. *backtest: Exploring Portfolio-Based Conjectures About Financial Instruments.* Version 0.3-4. R package. CRAN. `https://cran.r-project.org/web/packages/backtest/index.html`.

Goebelbecker, Eric. 2023. *Intro to Backtesting in R: A Comprehensive Tutorial.* Accessed September 8, 2025. `https://pipekit.io/blog/intro-to-backtesting-in-r-a-comprehensive-tutorial`.

Goodrich, Ben. 2024. *rstan: R Interface to Stan.* Version 2.32.7. R package. CRAN. `https://cran.r-project.org/web/packages/rstan/index.html`.

Grolemund, Garrett. *Hand-On Programming with R.* Sebastopol, CA: O'Reilly Media, Inc., 2014.

Grosser, Malte, Henning Bumann, and Hadley Wickham. *Advanced R Solutions.* Boca Raton, FL: CRC Press, 2021.

Guo, Xin, Lai Leung, Tze, Howard Shek, and Samuel Po-Shing Wong. *Quantitative Trading: Algorithms, Analytics, Data, Models, Optimization.* Boca Raton, FL: CRC Press, 2017.

Hanson, Daniel. *Learning Modern C++ for Finance: Foundations for Quantitative Programming.* Sebastopol, CA: O'Reilly Media, Inc., 2025.

Hiebeler, David E. *R and MATLAB®.* Boca Raton, FL: CRC Press, 2015.

Hvitfeldt, Emil. 2022. *lime: Local Interpretable Model-Agnostic Explanations.* Version 0.5.3. R package. CRAN. `https://cran.r-project.org/web/packages/lime/index.html`.

IBKRCampus, Traders' Academy. "Installing R and RStudio." Accessed September 2, 2025. `https://interactivebrokers.com/campus/trading-lessons/installing-r-and-rstudio/`.

IBKRCampus, Traders' Academy. "Introduction to R and RStudio." Accessed September 2, 2025. `https://www.interactivebrokers.com/campus/trading-lessons/introduction-to-r-and-rstudio/`.

Ihaka, Ross, and Robert Gentleman. "R: A Language for Data Analysis and Graphics." *Journal of Computational and Graphical Statistics* 5, no. 3 (1996): 299–314. `https://www.stat.auckland.ac.nz/~ihaka/downloads/R-paper.pdf`.

Interactive Brokers. "IBKR Order Types, Algos and Tools." Accessed September 28, 2025. `https://www.interactivebrokers.com/en/trading/ordertypes.php/`.

Josuttis, Nicolai M. *The C++ Standard Library: A Tutorial and Reference.* 2nd Edition. Upper Saddle River, NJ: Addison-Wesley, 2012.

Kaeppel, J. *The Option Trader's Guide to Probability, Volatility, and Timing.* New York: John Wiley & Sons, Inc., 2002.

Kalinowski, Tomasz. 2025. *keras: R Interface to 'Keras'.* Version 2.16.0. R package. CRAN. `https://cran.r-project.org/web/packages/keras/index.html`.

Kalinowski, Tomasz. 2025. *tensorflow: R Interface to 'TensorFlow'.* Version 2.20.0. R package. CRAN. `https://cran.r-project.org/web/packages/tensorflow/index.html`.

Kuhn, Max. 2024. *caret: Classification and Regression Training.* Version 7.0-1. R package. CRAN. `https://cran.r-project.org/web/packages/caret/index.html`.

Kuhn, Max. 2025. *tidymodels: Easily Install and Load the 'Tidymodels' Packages.* Version 1.3.0. R package. CRAN. `https://cran.r-project.org/web/packages/tidymodels/index.html`.

Lamb, James. 2025. *lightgbm: Light Gradient Boosting Machine.* Version 4.6.0. R package. CRAN. `https://cran.r-project.org/web/packages/lightgbm/index.html`.

Liaw, Randy. 2024. *randomForest: Breiman and Cutlers Random Forests for Classification and Regression.* Version 4.7-1.2. R package. CRAN. `https://cran.r-project.org/web/packages/randomForest/index.html`.

Lippman, Stanley B., Josée Lajoie, and Barbara E. Moo. *C++ Primer.* 5th Edition. Upper Saddle River, NJ: Addison-Wesley, 2013.

Lopez-Lira, Alejandro. *The Predictive Edge: Outsmart the Market Using Generative AI and ChatGPT in Financial Forecasting.* Hoboken, NJ: John Wiley & Sons, Inc., 2024.

Masters, Timothy. *Testing and Tunning Market Trading Systems: Algorithms in C++.* New York, NY: APRESS, 2018.

Matsuura, Kentaro. *Bayesian Statistical Modeling with Stan, R, and Python.* Singapore: Springer, 2022.

Meyer, David. 2024. *e1071: Misc Functions of the Department of Statistics, Probability Theory Group.* Version 1.7-16. R package. CRAN. `https://cran.r-project.org/web/packages/e1071/index.html`.

Meyers, Scott. *Effective C++: 55 Specific Ways to Improve Your Programs and Designs.* 3rd ed. Boston, MA: Pearson Education, 2008.

Miller, Eli. 2025. *riskmetric: Risk Metrics to Evaluating R Packages.* Version 0.4-1. R package. CRAN

Thomsett, Michael C. *Advanced Options.* Singapore: John Wiley & Sons, 2014.

Voellenklee, Andreas. 2020. "candlesticks: Candlestick Pattern Recognition." Accessed January 26, 2025. `https://rdrr.io/rforge/candlesticks/`.

Zakamulin, Veleriy. 2017. *Market Timing with Moving Averages: The Anatomy and Performance of Trading Rules.* Cham, Switzerland: Pallgrave, Macmillan.

Zakamulin, Veleriy, Javier Giner. 2025. *The Ultimate Moving Average Handbook: Bringing Science into the Art of Trend Following.* Cham, Switzerland: Pallgrave, Macmillan.

Index

For Product Safety Concerns and Information please contact our EU representative GPSR@taylorandfrancis.com Taylor & Francis Verlag GmbH, Kaufingerstraße 24, 80331 München, Germany

Batch number: 10722100

Printed by Printforce, the Netherlands